COAL TOWNS

OF WEST VIRGINIA VOLUME TWO

MARY STEVENSON

Charleston, West Virginia

First edition
Printed in the United States of America

10 9 8 7 6 5 4 3 2

Library of Congress Catalog Card Number: 98-066203
ISBN-13: 978-1-891852-28-2
ISBN-10: 1-891852-28-0

Book and cover design: Mark S. Phillips

Distributed by:

West Virginia Book Company
1125 Central Avenue
Charleston, WV 25302

www.wvbookco.com

TABLE OF CONTENTS

ACKNOWLEDGEMENTS

My personal journey to preserve history came by accident. Thomas Gales, Petroleum Engineer, gave me a box of old photographs and negatives of coal towns dating from the early years of the twentieth century. The pictures left a trail of the way life was in the early days of the mining of coal in the West Virginia mountains. This picture of how our grandparents worked and lived is not glamorous, but it is a picture of a hard-working generation of men and women.

My thanks to all the friends who helped me with *Coal Towns of West Virginia Volume 2*: Rebecca Black, Lois Moore, Maude Skiles, Robert Bodnar, Leona Wilcox, Margaret Smith, Francis Miller, Robert Hancock, Jack Roop, Ralph Wells, C. H. Wells, Mr. & Mrs. Robert Moore, Jeannie Kincaid, and Winnie Shumate.

Credits have been identified in this book with the photos and I certainly thank each and every one of those persons. I am very grateful to Ruth Hedrick Tully who gave me most of the wonderful pictures of Beckley. Thanks to Jessica and Nicholas Legg who assisted me with selecting and reproducing the pictures and for invaluable help on the computer. Thanks to Josephine Moore for reading my manuscript before publication.

I am grateful to Beckley Newspapers Inc. for the use of their library and permission to use some of their pictures. When undertaking a project such as this, it is so easy to omit someone who should have been given credit for their help. I am grateful to the many people who have just taken time to talk to me and answer my questions.

Lastly, I am grateful to my husband, Dan, for his support, encouragement, and company as we traveled to gather information for this book.

INTRODUCTION

Over millions of years coal was formed deep in our mountains. Much of our West Virginia history begins as man developed ingenious ways to mine this coal. The coal and the timber industries have played a tremendous part in the development of southern West Virginia, providing enormous wealth to a relatively small number.

This book is a combination of my first two pictorial histories, *From Affinity to Winding Gulf* (1989) and *From Ameagle to Wingrove* (1990). My third book, *Coal Towns of West Virginia* (1998) covers some of these communities but stands alone, as the pictures are not duplicated in either of these books. In this latest combination of *Affinity* and *Ameagle* books, *Coal Towns of West Virginia Volume 2*, I have included new pictures of Beckley and of several other surrounding communities.

As it developed, Beckley was given the unofficial title of "Capitol of the Coal Fields." The city is situated in the midst of one of the greatest coalfields of high quality coal in the United States. Beckley became the trading center for dozens of nearby coal mines. While it was necessary for the coal companies to provide housing for their miners near the mines, the mine owners, land development companies, and sales offices wanted to be located in a more convenient location. Beckley was the county seat and the logical place to establish the majority of these offices. Banks were formed. Lawyers, engineers, and bookkeepers were needed to run coal operations. They were brought to Beckley.

At one time coal, directly or indirectly, supplied the merchant, the professional man, the banker, the manufacturer, the distributor, and the laborer with most of the state's work, wages, and industry.

Many changes have occurred over the years. Coal companies eventually closed their mines and moved on, followed by the company store and post office. Families were forced to relocate, wherever there was work. Little remains of some of the coal towns except fragments of a tipple or a few stone chimneys.

The more I read about the men who opened and worked our coal fields, the more I stand in awe of their strength. I see over and over the names of the first coal barons who started out as miners and advanced to foreman or owners. The names of Collins, Dixon, E. E. White, John and James Laing, Gaston & George Caperton, and Tams are so intertwined that it is difficult to think of any coal company that these men did not help to grow.

However, we cannot think of the success of these leaders without considering the hundreds of thousands of miners who gave up their lives in the coalfields. Without the young and old men who crawled into the dark cavities of the earth, the success of the coal companies would not have been possible. But it also required the work of the women who made homes wherever the coalfields drew their husbands. It naturally led to include the children who grew up to either dig coal or go out in the world to become doctors, lawyers, teachers, and world leaders. All were necessary to make the picture complete.

I have the utmost respect for those men who, in order to provide for their families, worked deep in the earth, enduring dark, damp, and the dangers of hidden gas. Their women were a strong breed who lived without indoor plumbing, electricity, and most modern conveniences. Cooking on a coal stove by the light of a kerosene lantern, washing clothes on a washboard, and drying clothes on a fence were difficult. But even more so, was the daily waiting and worrying they endured until their husbands, fathers, and sons returned from the mine.

In talking to and reading about this early group of pioneering miners, they most often reported that life was good in the coalfields. There was a close community feeling and your neighbors were your friends. They were there for each other in good times and bad.

With the passing of time, we only have old pictures and fading memories to record this fascinating time and place.

FAYETTE COUNTY

The Gauley and New Rivers, along with the new C & O Railway, brought the coal operators and financiers to Fayette County several years before Raleigh or Wyoming Counties. Some mines were opened in the 1870s at Fire Creek, Quinnimont, Hawks Nest and Caperton. Colonel George W. Imboden bought a large tract of land from James Westlake for David T. Ansted, which formed the Gauley-Kanawha Coal Company in 1873. In 1875 it was re-organized and became the Hawks Nest Coal Company, Limited.

William H. Holland, John Nuttall, William N. Page were among the many who opened productive coal mines in this relative wilderness.

One of the outstanding men to come to this area was Colonel Joseph L. Beury, member and manager of the first coal company organized in the great New River coal field known as the New River Coal Company. This was the first company to ship the celebrated New River coal with the first shipment made in 1873. His association with his Quinnimont mine gave the town that name. He opened up various mines at Ansted, Caperton and Beury.

Thomas Gaylord McKell was an early pioneer in the Fayette County coal fields. But his interests were varied. In 1896 he built the Opera House at Glen Jean, and in 1901 he built the Dun Glen Hotel at Thurmond. He was the builder of the Kanawha, Glen Jean & Eastern Railway. Among his mines were Derryhale, Kilsyth, Oswald, Graham, and Tamroy.

After the death of Thomas in 1904, William McKell took over the management of his father's properties. He organized the Bank of Glen Jean in 1909 and served as president. He continued to expand on organizing mines and extended the length of the Kanawha, Glen Jean & Eastern Railroad.

Another early pioneer was Justus Collins, president of Collins Colliery. Leasing land from Thomas McKell, he opened several mines in the area, including Dun Loup, Turkey Knob, and Whipple. Upon selling his mines in Fayette County he took out leases on land in Winding Gulf and organized and opened mines under the name of Winding Gulf Collieries.

Probably Fayette County's most famous coal baron was Samuel Dixon, born in Normandy, England in 1855. Dixon came to West Virginia as a very young man, starting his climb to success as a bookkeeper for his uncle, Frederick Faulkner. Dixon served with various companies before combining his skills with Phineas W. Sprague to form the New River Company. It was reported that Sam Dixon put most of the earnings of his companies from 1893-04 into other properties in order to form the New River Company.

Under Dixon's leadership, the New River Company grew and expanded by purchasing more mines. He also built the White Oak Railroad to open up mines in Wingrove, Scarbro, Whipple, Oakwood, Carlisle, Summerlee, and Lochgelly. Eventually, the New River Company controlled most of the New River coal field.

RALEIGH COUNTY

A stage line ran its first coach between Lewisburg and Charleston in January 1827. Within ten years stagecoaches were operating on a regular schedule, carrying passengers and mail between Richmond and Guyandotte in four and one half days.

Coal and timber would have to be considered the bloodline to the growth of Raleigh County. With the arrival of the railways, timber and coal could reach distant markets and foreign ports.

In 1890 two Scotsmen, James Kay and John Laing, formed the Royal Coal & Coke Company and opened a small drift mouth mine high up on the mountainside on the New River in Raleigh County. Buckets transported the coals from the drift mouth across New River to the C & O main line on the other side of the river.

By 1900 the C & O Railway Piney Creek branch arrived in Raleigh County, making the way for other coal mines to be opened. As tracks were built into new areas each year, new companies were formed and began operating mines in great numbers.

Small lumber mills had operated for domestic use but with the approach of the iron rails, big companies started to form. Among the first were the Beaty Lumber Company, Blue Jay Lumber Company, George Bair Lumber Company, and W. M. Ritter Lumber Company.

150 Years

The Raleigh County Commission
cordially invites you to attend the
Sesquicentennial Birthday Celebration
of Raleigh County
to be held
Sunday, January 23, 2000
from the hours of
2:00 p.m. to 6:00 p.m.
at the
Raleigh County Courthouse
Beckley, West Virginia

1850
January 23rd
2000

Sesquicentennial Birthday Celebration – Raleigh County – January 23, 2000.

RALEIGH COUNTY CENTENNIAL 1950

Wooden nickels and the Raleigh County Centennial Seal.
August 27- September 2, 1950.

WYOMING COUNTY

In Wyoming County in the 1890s, Mr. Henry Huttleston Rogers organized the Virginian Railway Company by combining the five-mile lumber-carrying Deepwater Railway on the Kanawha with the Tidewater line of Eastern Virginia. He planned to build a railway on the most direct route from Norfolk, Virginia, to the coal fields of the Kanawha and New River-Pocahontas fields.

Mr. Rogers organized the Virginia Railway Company and took deeds for right of way through a portion of Wyoming County in July 1907. The construction started soon after and passed through Mullens and Maben, thereby opening the coal business in upper Wyoming County and the Ritter Lumber Company operations at Maben. Extensions of this railway were built into the Winding Gulf coal field to Itmann. In the late 1920s an extension down Guyandotte River to Gilbert made possible the opening of modern coal installations at Wyoming and Marianna. The Winding Gulf Branch included 24 miles of rugged mountain terrain, requiring three tunnels, to open up the rich coal fields.

Mr. Rogers died shortly after the completion of his railroad in 1909. His railroad allowed many coal operators who entered the rough mountain terrain of Wyoming County to make their fortune.

The first coal operator to begin operations was J. C. Sullivan, who located at Mullens around 1912 and opened a mine on the hill above the public road at South Mullens, at Mead Poca. The tipple was a complicated affair, sloping downward above the road and across the river where the coal was dumped into railway cars. It continued in business until the general slump following World War I, when it was the first local operation to close. It was dismantled about 1924.

Mr. Sullivan was also one of the owners and managers of Harty Coal Company at Tralee, which went into operation on a lease acquired in 1915.

Mr. Sullivan actively promoted development and progress of the town of Mullens. He joined in establishing the *Mullens Advocate*, and was promoter and part owner of the Wyoming Hotel. He helped to re-organize the Bank of Wyoming, which was moved from Oceana to Mullens in 1915. He also helped to found Wyoming County's first Catholic Church.

Many companies were to follow the black veins in the earth, giving work to men of many nationalities. Alpha Pocahontas Coal Company, 1915; Cooper Pocahontas Coal Company, 1915; Iroquois Coal Mining Company, 1915; Amigo Coal Company, 1915; Devil's Fork Coal Company, 1915; United Pocahontas Coal Company, 1915; Harty Coal Company, 1915; Zenith Coal & Coke Company, 1915; and Thermo Pocahontas Coal Company, 1916.

As World War I created a great demand for coal, Wyoming Land Company issued leases to Virginia Smokeless Fuel Company in 1914, Savine Smokeless Coal Company in 1914, and Sabine Collieries in 1917.

Coal operator Isaac T. Mann opened the Pocahontas Fuel Company in 1918. Mr. Mann named his town and mine Itmann, which was located about four miles from Mullens.

C. H. Mead, who served as president, organized Faith Smokeless Coal Company in 1925. Mr. H. H. Rogers and G. W. Hyman, a Boston financier, were the organizers of Glen Rogers and the Raleigh Wyoming Coal Company.

AFFINITY

General view of Affinity.

Affinity – Raleigh County: Pemberton Coal & Coke Company began operations in 1910. The report from the *1912 W.V. Dept. of Mines* stated: This is a slope and shaft mine, located at Affinity on Soak Creek, operating the Beckley seam, opened out on the triple and double entry system; average thickness of vein, four feet; good roof. Mining done exclusively by mining machines. Some sections of the mine get dry and dusty but are promptly watered down. A 10-foot Jeffrey Force fan produced the ventilation. Produced 53,907 tons of coal in 1912. The 1919 report listed 156,044 tons that year.

Pemberton Coal & Coke Office & Store.

AFFINITY

View of miner's house.

Superintendent's home and amusement hall.

ALPOCA

General view.

Alpoca – Wyoming County: Named for Alpha Pocahontas Coal Company, coal lease granted 1915 and produced yearly tonnage in 1919 of 114,146 tons.

Store and office building.

AMEAGLE

Ameagle - Raleigh County: American Eagle Colliery Company operated a drift opening at Ameagle. H. H. Howell was the mine foreman. In 1920 the American Eagle Colliery Co. employed 36 people inside and 20 laborers outside and worked an average of 130 days that year. No coke ovens were running during that year. They produced 22,281 tons of coal and furnished local trade and tenants with 214 tons, shipping the remainder by C & O Railway Company. Ameagle is on the Clear Fork of Coal River.

In the 1930s, the American Eagle Colliery Company employed 600 people. With an annual payroll of approximately $800,000, it was a sizeable contributor to the welfare and prosperity of Raleigh County. Ameagle, with a population around 3,000 people, had modern homes, churches, and a large company store.

PHOTOGRAPHS OF AMEAGLE COURTESY HORACE ROBINSON

Town of Ameagle 1950.

AMEAGLE

Town of Ameagle.

Ameagle cleaning plant tipple, 1950.

AMIGO

Amigo - Raleigh and Wyoming Counties: Amigo was named for the Amigo Smokeless Coal Company, which was established in 1914 by Dr. Joseph A. Wood. Amigo produced low-volatile coal that produced a very low ash content and was in great demand. In 1945 Dr. Wood's sons sold the company to E. C. Minter Coal Company. Still later Lillybrook Coal Company purchased it.

View of Amigo.

AMIGO

Store, offices, and restaurant at Amigo Coal Company.

Store and railway station.

BECKLEY

Beckley is situated in the midst of the greatest smokeless coal fields in the United States. The county seat of Raleigh County, Beckley is located about 2,500 feet above sea level and on top of a broad plateau of the Allegheny Mountains. Beckley became the trading center for over a hundred nearby coal operations of the great Winding Gulf, New River and other coal seams.

In a 1925 promotional pamphlet it was reported that Beckley "was modern in every respect with every desirable attribute for the home builder, and a clean intelligent citizenship, with all the big city conveniences, but without their congestion and other bad features. Beckley is an ideal residential, educational, religious, business, and healthful [sic] city."

Beckley's early growth had a direct correlation to the area's surrounding coal mines. The city's bankers, lawyers, accountants, builders, doctors, and merchants were all closely connected to the coal industry. The owners of the coal companies became stockholders of the banks, the local businesses, and helped to start many of the local churches.

The 1921 telephone directory contained listings for nine coal companies and shippers in the City of Beckley. They were Beaver Coal Company, C. H. Mead Coal Company, Bailey-Wood Coal Company, Rock Lick Coal Company, Ragland Coal Company, Meadow River Smokeless Coal Company, Hawley Coal Company, Raleigh Smokeless Fuel Company, and E. C. Minter Coal Company.

- The coal companies brought men such as Col. Ernest Chilson, general superintendent and president of Raleigh Coal & Coke Company and Knights Products Company.

- Thomas H. Wickham was president of Beaver Coal Company, Raleigh Smokeless Fuel Company, and Lillybrook Coal Company. He was vice president of the Bank of Raleigh; he was active in building, real estate, and insurance. He served as Mayor of Beckley, finishing the term of Lucian H. Davis. Wickham was a dedicated worker in the Presbyterian Church.

- Halleck McGinnis Scott, born in Crow, WV, returned to Beckley after his schooling with a degree in Civil Engineering. He became associated with the Winding Gulf operations.

- Samuel E. Turner was part owner of Smokeless Coal Company in Winona. Selling his shares, he moved to Beckley and established the Rose and Turner Furniture Company with partner C.C. Rose. Turner started another furniture store on Main Street with Burnell Ware.

- Roy J. Burmeister was connected with local businesses and the New River Collieries Company at Eccles.

- Lorin T. Putman was general manager of the Raleigh-Wyoming Company, with properties at Edwight and Glen Rogers.

- John Wallace McCreery was a popular lawyer working with coal companies, banks and a wide variety of local businesses. He was a state senator for several terms and the founding father of the Beckley Presbyterian Church. Senator McCreery was probably one of the most popular Beckley citizens and was honored by having McCreery Street named for him.

- John Baxter Earwood was a lumberman and realtor, buying, selling or cutting large tracts of timber throughout this region. One of his projects was the huge forest in Richmond District, between Shady Spring and Grandview. He shipped his lumber over the C & O Railway branch between Beckley and Prince.

- J. A. Hunt, president of Lillybrook Coal Company, Raleigh Smokeless Fuel Company and the New River Company.

- Thomas K. Laing, whose father had opened Royal mine in 1892, was a lawyer and entered the coal production field at Willis Branch.

- J. B. Clifton, president of Raleigh Smokeless Fuel Company and Lillybrook Coal Company.

- Ashton File, attorney and director in many local coal companies, including Lillybrook, Raleigh Smokeless, Red Ash, and Gulf Mining Company.

- J. Lewis Bumgardner, attorney who organized numerous coal operations including East Gulf Coal Company. He was active in civic affairs and the Presbyterian Church.

- John R. Hornbrook, who with Prince Lilly, organized Lillybrook Coal Company, Lilly & Hornbrook and later Gulf Mining Company.

- E. E. White, founder, organizer, and president of E. E. White Coal Company at Glen White and Stotesbury.

- Major W. P. Tams, Jr., president of Gulf Smokeless Coal Company at Tams.

- E. C. Minter led operations of Beckley Smokeless, E. C. Minter Coal Company, Minter Fuel Coal Company and many others.

As Beckley grew, hospitals were built to serve the needs of the citizens, as well as banks, offices, and other new businesses. Many of Beckley's fine doctors arrived in the area after being hired by the coal companies. Dr. J. A. Campbell, a company doctor for Raleigh Coal & Coke, opened the first hospital on Neville Street, and Dr. W. W. Hume started a small hospital in Mabscott that was later moved to Beckley. Dr. Campbell established a training school for nurses in 1910.

With the growth of population came the growth of public utilities, bringing the telephone to Raleigh County in 1901. Electric lights came to the courthouse in 1905. The wells, springs, and cisterns were replaced with the first piped water in November 1907, followed by a municipal sewer system in 1908. The main business area's streets were bricked in 1910. With such modern public services the city continued to grow under the guidance and leadership of dedicated men. The town of Beckley became the City of Beckley in 1908.

In *Beckley USA*, Harlow Warren stated that the early records of Beckley mayors were not complete, but that John W. McCreery had served prior to 1897 and Capt. John Anderson was mayor in 1898-1899. After Beckley became a city, Lucian H. Davis was elected for the 1910-1911 term. He resigned in midterm and was replaced by Thomas H. Wickham.

Entertainment soon arrived in Beckley. Fairyland Theatre opened on East Main Street, Wonderland Theatre opened on Heber Street, and the Carter Opera House and the Lyric Theatre opened on South Fayette Street. The Palace Theatre opened in 1923 on North Kanawha Street with W. A. Burke as manager. The Beckley Theatre opened in 1936.

As the roads and cars improved, many of the owners and superintendents moved their families to Beckley. It would have been difficult to think of any business in Beckley that did not benefit from the coal industry.

The payroll of the mines constituted the base of the city's economic life. The merchants, bankers, traders, and professional men of the city were dependent upon them for business.

Lacy A. Dillon, in his book "*They Died For King Coal,*" stated that Beckley was the only city in southern West Virginia to suffer a mine explosion inside its corporate limits.

On December 17, 1940 a gas and dust explosion in the Raleigh No. 4 mine killed seven men, injuring four others. Two later died in the hospital bringing the total to nine dead. The No. 4 mine, owned by the Raleigh Coal and Coke Company, had been bored into the plateau below Woodlawn Avenue, one of the best residential neighborhoods in the city.

The City of Beckley was built on the rolling and somewhat level terrain of plateau Allegheny Mountains and was atop countless mines in the earth below.

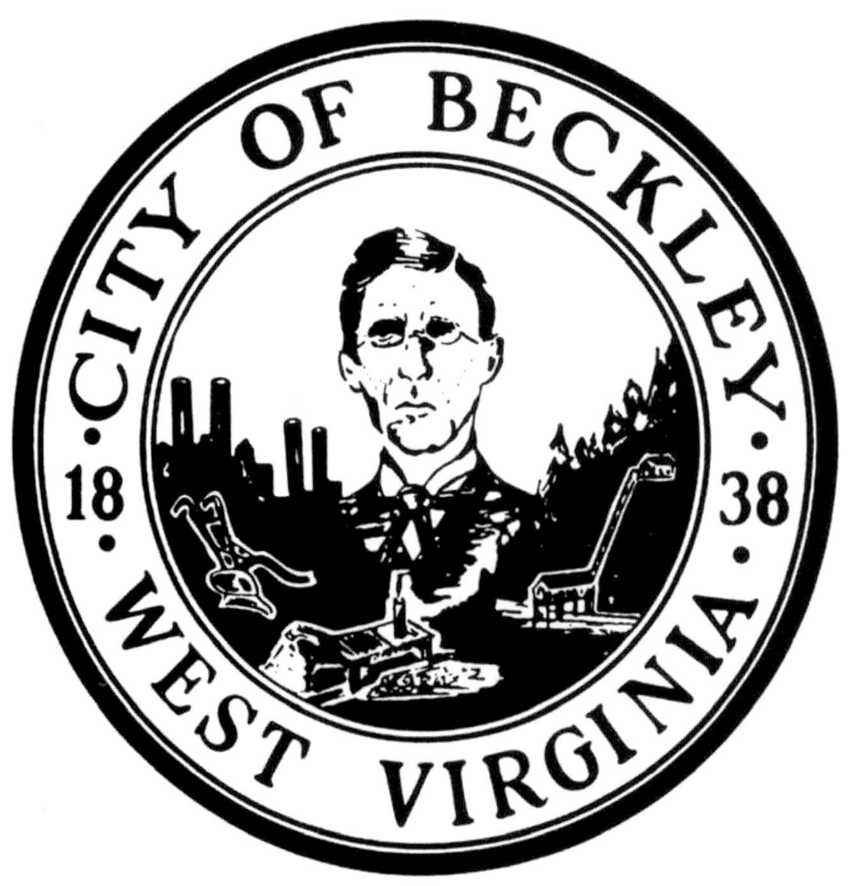

Corner of Main Street looking down South Kanawha Street.

The road sign at the left states this is State Routes 12 and 65. Mullens - 27 Miles and Harper - 6 miles. To the south Hinton - 25 miles and Bluefield - 50 miles. North to Fayetteville – 24 miles and Charleston – 75 miles.

The small brick building to the left is the law office of McCreery and Patterson.
COURTESY CRYSTAL PATTERSON

Dr. E. S. (Lillian) Dupuy residence on South Kanawha Street.
Mrs. Dupuy was a daughter of Samuel Dixon of the New River Company.

Home of Mr. and Mrs. Harry H. Pinckney.

This beautiful home is located on South Kanawha Street. Mrs. Pinckney and Mrs. Lillian Dupuy were sisters and lived next door to each other. They are both daughters of Samuel Dixon of the New River Company. Mrs. Dupuy's home is pictured on the previous page. Mrs. Harry M. (Crystal) Patterson still resides in this house.

Harry H. Pinckney was the manager of the Skelton commissary for a few years.

The Beaver Coal Company bought this house, moved it to the back of the lot, and built their office building in front of the house. Mr. William MacTaggart was superintendent of Beaver Coal Company for many years and was succeeded by Lucien H. Davis.
COURTESY CRYSTAL PATTERSON

White Cross Pharmacy, Drugs, Sundries, and Stationery
in their first location in G. C. Hedrick's property on Main Street.

Magnesium safe of the Beckley National Bank
costing $2,550.00
COURTESY RUTH HEDRICK TULLY

Beckley Hardware and Supply Company.

Beckley Hardware and Supply Company, established in 1909, was located in a small frame building on Main Street that was destroyed by the fire of April 1912. Upon completion of a new building owned by Mrs. E. Z. Burgess Grubbs on Heber Street at the foot of Main Street, the corporation leased the property and occupied it late in 1912. The stockholders were G. C. Hedrick, E. P. Ewart, C. L. Beckner, S. E. Turner, W. I. Ford, W. H. Rardin, J. A. Ewart, B. E. Carter, and T. E. Bibb. COURTESY RUTH HEDRICK TULLY

BECKLEY

This was the second Raleigh County Courthouse that was built on land deeded to the City of Beckley by General Alfred Beckley. The brick construction was completed on July 6, 1893 at a cost of $34,354. The building was used until 1935 when the current building was erected. COURTESY BECKLEY NEWSPAPERS

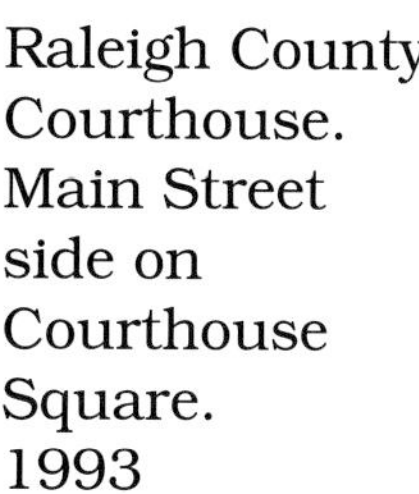

Raleigh County Courthouse. Main Street side on Courthouse Square. 1993

Beckley Presbyterian Church – (1924) - South Kanawha Street
J. L. Lineweaver standing on steps.

Home of Mr. and Mrs. P. H. M. (Willie) Patterson

This house was sold to the Presbyterian Church to make way for their Educational Building in 1954. COURTESY CRYSTAL PATTERSON

P. H. M. Patterson, Sr., in front of the McCreery & Patterson Law office.

This small brick building stood on the corner of Main and Kanawha Streets for many years and was truly a landmark. It housed the law firm of Senator John W. McCreery and his son-in-law, P. M. H. Patterson, Sr. COURTESY CRYSTAL PATTERSON

Raleigh County Bank.

Grover C. Hedrick was the founder of Raleigh County Bank, organized in 1909 with a capital stock of $25,000. Among the stockholders were Boyd E. Carter, W. H. Rardin, J. E. Summerfield, T. E. Combs, Robert Wriston, A. B. Clay, George W. Bair, R. T. Bair, and J. Lewis Bumgardner. Raleigh County Bank absorbed Peoples Bank. It is now the United National Bank.

Mr. Hedrick was twice mayor of the City of Beckley, banker, business promoter and one of the outstanding community builders of Raleigh County. He was director and stockholder in many coal, utility, land, farming and real estate corporations.
COURTESY RUTH HEDRICK TULLY

Dr. J. E. Coleman and Dr. Robert Wriston opened the Beckley Hospital on South Oakwood Avenue in 1913. Dr. J. H. McCulloch of Charleston and Dr. A. U. Tieche of Winding Gulf purchased this hospital in 1922. They built a new hospital in 1929 and turned the above building into a nurse's home.

Dr. W. W. Hume's home on North Kanawha Street.

Dr. W. W. Hume started a small hospital in Mabscott, later moving it to Beckley on North Heber Street. Dr. Hume was actively involved with several business ventures in the area. Raleigh County Medical Society was formed in 1903 with Dr. Hume as its first president. COURTESY BECKLEY NEWSPAPERS

George W. Bair home on North Kanawha Street.
Bair was a lumberman, builder, and car dealer.

The Bair home is currently the home of Concord College, Beckley Branch.

George H. Moss home, North Kanawha Street.
Founder of Raleigh Hardware Company.

Home of Charles L. Allen, Civil Mining Engineer, South Kanawha Street.
Current location of Beckley Municipal Building.
COURTESY JULIA L. WILLSE

Residence of John W. McCreery on South Kanawha Street.

Senator McCreery was a banker, coal operator, realtor, lawyer, editor, and prosecuting attorney. He was the first president of the Bank of Raleigh and founding father of the Beckley Presbyterian Church. McCreery donated the lot across from his house to his parish for their first church.

P. M. H. Patterson, Sr., and Senator John W. McCreery in front of their law office. This brick building was on the corner of South Kanawha and East Main Streets where the City National Bank is now located. COURTESY CRYSTAL PATTERSON

The stately home of Joe L. Smith, Neville Street.
Birthplace of Governor Hulett Carlson Smith.

Congressman Joe L. Smith was owner and publisher of the *Raleigh Register*, president of Beckley National Bank, and a dedicated civic leader. Mr. and Mrs. Smith lived in this house several years before building a beautiful stone house on Woodlawn Avenue.

Joe L. Smith is the father of Hulett Carlson Smith. Hulett Smith was born and grew up in Beckley, becoming the twenty-seventh Governor of West Virginia. Hulett, graduating from the Wharton School of Finance and Administration with honors, returned to Beckley to enter the insurance business. In 1948 he was selected "Outstanding Young Man of the Year" by the Jaycees. Of his political career he remarked, "The door is gradually closing, but nothing is ever completely shut."

Both Joe and Hulett Smith represented West Virginia in Congress in Washington.

Beckley Hotel corner of Main and Kanawha Streets.
COURTESY BECKLEY NEWSPAPERS

Rose Funeral Home.
COURTESY BECKLEY NEWSPAPERS

Home of Dr. and Mrs. Robert Wriston on Neville Street.
In 1913, Dr. Wriston helped organize and was half owner of Beckley Hospital.

Home of Mr. and Mrs. W. Wesley Wilkes on Neville Street.

Mr. Wilkes was the founder of Wilkes Insurance Agency, one of the largest insurance agencies in the state. The above home was previously owned by William McTaggart and had the first indoor bathroom in Beckley. Mr. and Mrs. Wilkes were the parents of Joan and Betty Wilkes.

BECKLEY

The Home of Dr. and Mrs. S. Monroe Bledsoe on North Kanawha Street.
COURTESY BECKLEY NEWSPAPERS

The Beckley Drug Store on Main Street was established by Carl Van Zandt with the assistance of Dr. S. Monroe Bledsoe, who later acquired the business and operated it for many years. COURTESY BECKLEY NEWSPAPERS

An unpaved Main Street.

Raleigh Motor Company-
East Main Street.
Bair Brothers - Buick & Ford
Cars.

J. Lewis Bumgardner's home on North Kanawha Street.
Attorney and organizer of East Gulf Coal Company.
COURTESY BECKLEY NEWSPAPERS

Dr. J. A. Campbell's hospital on Heber Street.

In 1905 Dr. Campbell opened a hospital on Neville Street near First Avenue, but the fire of 1912 destroyed it. He later he opened his hospital in this stone building on Heber Street around 1913.

Home of Emmett S. Pugh on Granville Avenue.

Mr. Pugh, well respected for his knowledge of bookkeeping, became actively involved in several coal companies: Raleigh Smokeless Fuel Company, of Lillybrook Coal Company, of Raven Coals, Inc., and the Raleigh Steamship Company. He was the grandfather of Beckley's current mayor, Emmett S. Pugh, III.

Mr. and Mrs. William McTaggart's Home on North Kanawha Street.

Mr. McTaggart was superintendent of Beaver Coal Company, and was associated with Raleigh Smokeless Fuel Company, Raven Ash, and Lillybrook Coal Company. This house was razed to build the North Kanawha Street post office.

Beckley Baking Company, South Oakwood Avenue.
"Aunt Betty Bread." COURTESY BECKLEY NEWSPAPERS

Lewisburg Ice Cream Company.
Located in Mabscott and was the forerunner of Greenbrier Dairy.

Old Beckley Armory on Park Avenue.
Home of two National Guard Companies.
COURTESY BECKLEY NEWSPAPERS

Gus Farris Department Store on Neville Street.
The department store was in business from 1911-1991 in three different locations.

The Bank of Raleigh, organized in 1899, was Beckley's first bank under the leadership of John W. McCreery as president, Isaac C. Prince, vice president, and E. Miller, cashier.
COURTESY BECKLEY NEWSPAPERS

Below: Interior of lobby.

Beckley Exchange Bank, corner of Main and Heber Streets.

Beckley National Bank

Winding Gulf Bank moved to Beckley in January 1914 with Dr. J. A. Wood as bank president. In July 1914 the name was changed to Beckley National Bank. They moved into their beautiful new building shown above in 1924. Joe L. Smith was bank president for many years, succeeded by Elmer L. Davis. COURTESY BECKLEY NEWSPAPERS

Central School Building

Central School Building was Beckley's first high school, built in 1918. The population of Beckley was soaring and within five years plans were made for a new high school. This building served the junior high children for several years and is currently a part of the Mountain State University campus.

Woodrow Wilson High School Building

The new high school, Woodrow Wilson, was built on land given to the Town District school board by the Beaver Land Company. It was completed in 1925 at a construction cost of $200,000. COURTESY BECKLEY NEWSPAPERS

John Earl McCreery shown here taking his sister, Miss Molly McCreery, and his young niece Elizabeth Johnson for what might have been a Sunday afternoon drive. The two ladies in the middle are unknown.

The house in the background was Senator John W. McCreery's home on South Kanawha Street, and the gentleman in the swing was D. Howe Johnson who married Pearl McCreery. After Senator McCreery's death, Howe and Pearl Johnson lived in this house. COURTESY CRYSTAL PATTERSON

BECKLEY

The Boyd Edwin Carter home on North Kanawha Street.
Carter-Walker House.

Mr. B. E. Carter and his bride came to Beckley in 1901. Mr. Carter promoted, if not operated, almost every conceivable type of enterprisc, and built over fifty business or dwelling structures. He built thirty houses at Sprague for the New River Company. He constructed the Carter Opera House on South Fayette Street, famous for its plays by local talent and traveling companies. He bought, sold, and leased many acres of valuable coal lands. With F. M. Cook he founded the Cook Carter Coal Company at Terry. He was a founder and president of Raleigh County Bank.

His major civic project was the promotion and development of the City Park on the 19-21 by-pass. The completion of this project called for thousands of dollars as well as months of grueling effort and political maneuvering, which he contributed freely. Harlow Warren said, "Few men, if any, have come to Beckley and worked harder, for more hours, than Mr. Carter."

This beautiful home built around 1901, was selected by the Raleigh County Landmark Commission to be listed on the "Raleigh County Historic Register of Historic Sites." The house is now called the Carter-Walker House.

Bair Building, corner of Main and Kanawha Streets.
COURTESY BECKLEY NEWSPAPERS

Palace Theatre, North Kanawha Street - built in 1923.
COURTESY BECKLEY NEWSPAPERS

BECKLEY

A postcard picture of Baptist Church on unpaved Neville Street. Writing on the postcard is dated 4-10-11.

Tombstone of Beckley's founding father, General Alfred Beckley, 1802-1888. Wildwood Cemetery.

Home of Mr. and Mrs. Thomas H. Wickham, North Kanawha Street.

The Wickham home was built in 1914 of native hand-cut stone. The original house was smaller and daughter Nancy laughingly remarked, "that every time she and Mama went away, they came home to find he had added another room." The beautiful house has 12 rooms plus three baths and is still occupied by daughter Nancy Wickham.

Mr. Wickham was the first superintendent of the Beaver Coal Company, and acquired large real estate holdings of his own. He was an early builder of houses on Wickham Avenue and built the houses that started the settlement of West Wickham. He served a partial term as mayor and was active in the Presbyterian Church.

Mr. and Mrs. William Robert Campbell's home on West Neville Street.

Mr. Campbell was a surveyor and for many years worked for Logan M. Bullitt, pioneering local coal development. He was sheriff of Raleigh County from 1905-1907. He had three daughters by his first wife: Wanda (Mrs. S. Monroe Bledsoe), Alma (Mrs. Ray J. Burmeister), and Ella (Mrs. E. Earl Bibb).

After the death of his first wife, he married Betty Beckley in 1894. Betty Beckley was the daughter of John Beckley and granddaughter of Alfred Beckley.

BECKLEY

STATION AT BECKLEY—RALEIGH COUNTY
Piney River and Paint Creek Railroad Co.

Train Station at Beckley – Raleigh County.
Piney River and Paint Creek Railway Company.
COURTESY J. W. DIXON, JR.

The Piney River & Paint Creek railway built a six-mile line from Cranberry to Beckley Junction in 1905-06, allowing passenger trains and mail service into Beckley. The first Beckley depot was built soon thereafter on Third Avenue.

The railroad depot at Beckley became busy. The Virginian Railway began passenger service to Beckley on the morning of December 16, 1912, unloading scores of passengers from Winding Gulf District. There was one mail and baggage car and two passenger coaches. The train made two round trips from Mullens on The Virginian daily, and was greeted on its first arrival by Beckley officials.

Chris Kambouris's Rest-a-While Lunch Room.
COURTESY BECKLEY NEWSPAPERS

Pete Joannides' DeLux Café.
COURTESY BECKLEY NEWSPAPERS

Boyles Cash Store in East Beckley.
COURTESY BECKLEY NEWSPAPERS

Interior View of Boyles Cash Store.
COURTESY BECKLEY NEWSPAPERS

Wood working plant of Byus-Mankin Lumber Company. North Kanawha Street. COURTESY BECKLEY NEWSPAPERS

View of Byus-Mankin Lumber Company's yard. COURTESY BECKLEY NEWSPAPERS

Samuel E. Turner's home on Beckley Avenue.

Mr. Turner was born in Ayshire, Scotland in 1866. The coal business brought him to West Virginia where he became a part owner of Smokeless Coal Company in Winona. He sold his interests and moved to Beckley to establish Rose and Turner Furniture Company. He later went into business with Burwell Ware. He was active in several civic organizations and elder of the Presbyterian Church. COURTESY RUTH HEDRICK TULLY

C. M. Calloway Residence on North Kanawha and Quesenberry Streets.

Charley M. Calloway owned the C. M. Calloway Lumber Mill located in Sprague Bottom. It later moved to high ground on North Kanawha Street. Calloway and Son sold to the Bair Brothers. The next owners were Charley Byus and Bob Mankin, hence the name of Byus-Mankin Lumber Company. After Mr. Calloway's death, the house was sold to E. C. Minter. COURTESY RUTH HEDRICK TULLY

Home of Burwell Ware on North Kanawha Street.

Mr. Ware was a well-known furniture dealer with Thompson & Ware and Turner & Ware. Dr. W. W. Hume and Burwell Ware are standing on the porch. COURTESY RUTH HEDRICK TULLY

Beautiful home of Robert T. Bair, South Kanawha Street.

The Bair home was built in 1905 for M. R. Matheny and was sold to Mr. Bair a few years later. Mrs. Bair's father, Samuel Kent, built the house for Mr. Matheny. Mr. Robert T. Bair was a lumberman, builder, and car dealer in businesses that he operated with his brother George W. Bair, Jr. It was reported that Bob and George brought the first Ford into Beckley in 1907. They opened a car dealership, receiving their first shipment of Fords in the spring of 1908. The vehicles were shipped to Beckley in boxcars and unloaded at Mabscott, where Mr. C. C. Farmer and a crew of men assembled them. The first four cars were sold to J. L. Bumgardener, B. E. Tolbert, H. H. Harwood, and Dr. E. L. Ellison. COURTESY THE BAIR FAMILY

The Home of Judge & Mrs. W. H. Rardin, 119 West Neville Street.
Rardin was a Raleigh County judge and vice president of Raleigh County Bank.
COURTESY RUTH HEDRICK TULLY

George W. Bair, Jr. Lumber Company, which produced interior finishes, and all kinds of mill work, brick, laths, coal, and lumber.

George W. Bair, Jr. came to Beckley at the age of sixteen, stayed a year, and then walked back to Tennessee to get his brother Bob. Both returned to Beckley on foot. George W. and Bob became two of Beckley's pioneer business and civic leaders. They started their lumber mill in 1871, located near the railroad tracks in Sprague Bottom.

COURTESY RUTH HEDRICK TULLY

Turner & Ware Furniture located at the corner of Main and Kanawha Streets.

Samuel E. Turner and Burwell Ware sold furniture and household goods. COURTESY RUTH HEDRICK TULLY

Interior of William Steiner's Jewelry Store on Main Street, circa 1908.

Steiner's was known as the leading jewelry store in Raleigh County with the latest fashions in jewelry. COURTESY RUTH HEDRICK TULLY

Beckley City Shop, J. L. Aliff, Manager.
Horse Shoeing & Repair Shop.
Man to the left with mustache is Thomas Floyd.
COURTESY JEANNE COLE

This banquet was given for D. C. (Lyle) Wade on his departure from the Raleigh County Bank to work for Slab Fork Coal Company. Shown clockwise: J. W. Dearing (The Fair), Boyd E. Carter (Raleigh County Bank), Cyrus H. Meador (later Beckley National Bank), Luther Cook (Ruffner Brothers), J. Garfield Foster (Foster Hardware Co.), Grover C. Hedrick (Raleigh County Bank), George Bair, Jr. (Bair Brothers), the honored guest, D. C. Wade, E. Earl Bibb (Beckley Hardware Co.), A. G. Meade (Meade Jewelry Co.), Robert D. Watts (R. D. Watts Insurance Co.), Carl Van Zandt (Beckley Drug Co.), Jesse Tolliver (Raleigh County Bank), Keith Davis (Raleigh Hardware Co.), John Mellen (building contractor), and S. Monroe Bledsoe (Beckley Drug Company). Restaurateur: Wilber Lloyd. The "Busy Bee" was later the Eatwell Café, owned by the Sarandon Brothers. *Beckley USA Vol. I* dated the picture 1912. COURTESY RUTH HEDRICK TULLY

Interior of the E. M. Payne Store on Main Street. Payne's for many years was "the place to shop," carrying good brands and offering excellent service. Payne Store opened in 1900 in a location at the corner of Main and Kanawha. The store pictured is on Main Street and in 1938 advertised, "Our stock is complete in every detail and the quality, style that you will see here speaks volumes in our favor." COURTESY RUTH HEDRICK TULLY

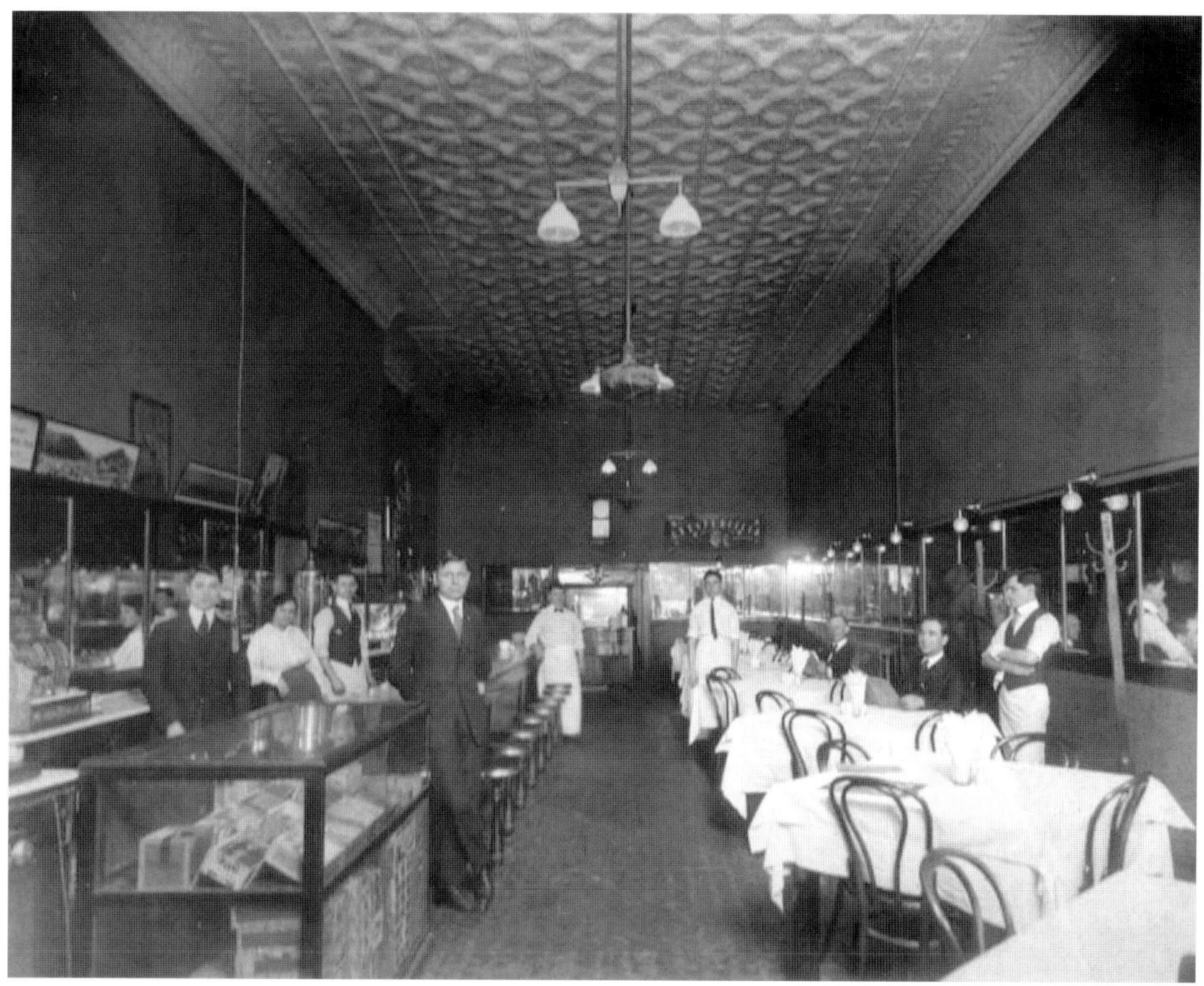

Only "A. Missus Restaurant" identified this picture with no date or place of business. It was personally autographed "To my really good friend Grover C. Hedrick, cashier of Raleigh County Bank. Friendship, A. Missus." A lovely restaurant, nice tables, neat servers, and a sign on the wall "Do Not Ask For Credit." COURTESY RUTH HEDRICK TULLY

BECKLEY

A bird's eye view of Beckley from Knob Hill on Harper Road.
COURTESY RUTH HEDRICK TULLY

Grover C. Hedrick, cashier of the Peoples National Bank.

The Raleigh County Bank under the leadership of Hedrick absorbed the Peoples Bank. Mr. Hedrick was mayor of the City of Beckley in 1932-33. Ruth Hedrick Tully is a daughter of Grover C. Hedrick. COURTESY RUTH HEDRICK TULLY

Peoples National Bank around 1910.

Left to right: Jesse Tolliver, Bob Bair, George Bair, Grover C. Hedrick, John Beckley, D. C. (Lyle) Wade, H. H. Harwood, and John Mellen. COURTESY RUTH HEDRICK TULLY

Corner of Main and Heber Street.
COURTESY RUTH HEDRICK TULLY

Interior of Raleigh County Bank (circa 1918) in their first building at 120 Main Street. B. J. Shumate is working in the center with a Mr. Kidd in background. COURTESY RUTH HEDRICK TULLY

The Oakwood Hotel on Oakwood Avenue after a snow storm.

The Oakwood Hotel across from the C & O Railway station was built and owned by Patrick L. Blankenship and was probably operated by his son, George Blankenship. Patrick L. Blankenship was sheriff of Raleigh County in 1921-24. COURTESY RUTH HEDRICK TULLY

First Christian Church – erected in 1918.

St. Frances de Sales Catholic Church.
Built in 1907, South Oakwood Avenue.
COURTESY BECKLEY NEWSPAPERS

Raleigh Presbyterian Church, South Kanawha Street.
Renamed Beckley Presbyterian Church in 1913.

In 1907 Senator John W. McCreery—the church's first member and first Elder—gave a lot on South Kanawha Street to build the church pictured above. The church was built of pressed brick at a cost of $8,000 and dedication services were held on July 25, 1909. As the church quickly outgrew the building, it was sold to the City of Beckley in 1927 for city offices and a fire department. A new church was erected on a lot very near the old structure. Dedication services for this new stone structure were held in October 1927. COURTESY RUTH HEDRICK TULLY

Methodist Temple, 500 South Kanawha Street.

The first Sunday services were held in the new Methodist Temple on February 5, 1928. The building was constructed by local contractor, E. F. Cooper, at a cost of $100,000, exclusive of furnishings. The church occupied the handsome building of stone and buff brick for over fifty years. This building is now part of Mountain State University.

St. Stephen's Episcopal Church, Woodlawn Avenue.

This brick church was built in 1927 after using the Methodist "Mother Church" since the turn of the century. In 2002, St. Stephen's celebrated their hundredth birthday. Mr. and Mrs. John B. Earwood with Mrs. Wirt French were the first Episcopalians in Beckley and organized the church. The first trustees were, J. B. Earwood, J. E. Summerfield, and Dr. J. A. Campbell. The J. B. Earwood's are the grandparents of Mrs. Robert (Margaret) Sayre and W. A. James, III.

First Methodist Church.

This was the second Methodist Church building, replacing one built of logs that stood on the corner of Neville and Heber Streets. The one above was located on Heber and Earwood Streets and was affectionately called the "Mother Church" because it was used by guest denominations in Beckley. The Methodists had split during the Civil War and the above was the north branch of the church. COURTESY RUTH HEDRICK TULLY

Methodist Episcopal Church South

General Alfred Beckley made a deed to the trustees of the Methodist Episcopal Church South in October 1859. General Beckley and his family were members of this church. The above building may have been built shortly before the Civil War. The building was at the corner of North Kanawha and Prince Streets. From the diary of Rutherford B. Hayes, he wrote on January 2, 1862: "The village of Raleigh is about ten to twelve years old; three or four hundred inhabitants may have lived there before the war; now about six or eight families. There were two churches, two taverns, and two stores in peaceful times. The troops lived comfortably in the vacated houses but [sic] too scattered."

This was possibly the second church built on this lot as records indicate that in 1902 a new church was built on North Kanawha and Prince Streets after "the little brown church was torn down." COURTESY RUTH HEDRICK TULLY

The Chesapeake & Ohio's Pullman car dedicated to the City of Beckley.

Left to right: Walter J. Trohy, President C & O Railway, Cleveland, Ohio, W. A. "Gene" Burke, mayor of Beckley, and Phil A. McDaniel, Beckley businessman.

The general public was invited to the ceremony on May 18, 1951 at the C & O tracks in Skelton where the locomotive sat for inspection. Two thirteen year olds from Beckley assisted with the celebration. At exactly 2:30 p.m. Ricky Thompson pulled the whistle cord on Engine "490," and Carolyn Morrison broke a bottle of champagne against the streamline coach. Phil A. McDaniel was master of ceremonies and the Woodrow Wilson Band furnished music. COURTESY FLO STEFFEY

City of Beckley Pullman car.

The dedication ceremony of the C & O Pullman car named to honor the city of Beckley on May 18, 1951, at the train tracks in Skelton. COURTESY FLO STEFFEY

Ricky Thompson blowing the locomotive whistles for the May 18, 1951 ceremony.
COURTESY FLO STEFFEY

Wildwood, home of General Alfred Beckley.

Log house now located at Wildwood, which was moved from Pax. The 1850 house was moved and reassembled to represent the post office that General Beckley once had located on his property.

Wildwood house library with many artifacts of the Beckley family.

The table in the foreground has a long history. The table belonged to John Beckley, father of Alfred Beckley. The story is that a part of the Declaration of Independence was signed on this table. More factual perhaps is that portions of the Bill of Rights may have been signed on this table as the document also bears his signature. John Beckley, at the time the Bill of Rights was signed, was the clerk of the House of Representatives. If John Beckley owned this desk at that time or later acquired it, the Bill of Rights could have been signed upon it. Years have passed and we probably will never really know the true story about the famed desk.

The desk was on display for several years in the Cultural Center of Charleston and was moved to Wildwood when it opened for tours. Dr. and Mrs. M. M. Ralsten bought the old home and with the assistance of the Raleigh County Historical Society restored it to as it was in General Beckley's day. Later the house was deeded to the City of Beckley to be open to visitors for tours. Mrs. Ralsten was a great-granddaughter of Alfred Beckley.

BESOCO

View of Besoco - Raleigh County: Beckley Smokeless Coal Company

Boarding house in Besoco.

E. C. Minter opened Beckley Smokeless Coal Company in 1915. He derived the name of his community from the initials of the company. The coal was shipped on the Virginian Railway.

BLUE JAY

Blue Jay - Raleigh County.
No. 4 Mine, 1909, Blue Jay Lumber Company.
COURTESY E. R. REYNOLDS COLLECTION

The 1920 Department of Mines reported Blue Jay No. 5 was a drift mine opening with P. J. McQuade as superintendent and mine foreman, working in six feet eight inch seams of coal. No. 5 produced 94,917 tons, used 210 tons in mining operations, and 540 tons for local trade and tenants. Blue Jay Lumber Company shipped on the C & O line.

BRADLEY – Raleigh County: Squire Jim Bailey, a farmer, settled on approximately 400 acres of farmland to raise a family of ten children. Later he built cabins to rent to travelers and opened a toll road across his property. He kept a team of mules handy in case anyone got stuck in the muddy road. He died in 1925 after being gored by a pet Guernsey bull while playing with the animal on the farm pasture. The countryside became know as Bailey's Cabins or the Bailey Farm.

Another early settler was Rufus Neal, whose stone house still stands near the ramp to Rt. 19. The first commercial business was a service station, operated by a grandson of Squire Jim Bailey.

Bradley, a farming community located about halfway between Beckley and Mount Hope, was never a coal mining town such as Sherwood, Tamroy, and Price Hill. But Bradley was close enough for men to live there and get to the work in the mines.

Samuel Dixon of the New River Company built the first grade school and leased it to the Board of Education. He also donated land for a cemetery to be used by the residents from his coal towns of Price Hill, Sherwood, and Tamroy. The Price Hill cemetery and Price Hill Elementary School were located in what is now Bradley. Later the Board of Education was petitioned by the school to change the name to Bradley Elementary.

Willis Stover served as principal of this school for many years. His wife, Dorothy, taught school at Oswald before coming to teach at Bradley. It is commonly thought that every child in Bradley for 45 years went to school with either Willis or Dorothy Stover. In speaking of local residents such as past Raleigh County Commissioner Vernon Barley, Dorothy is quick to point out that "I had Vernon in the first grade when I taught school in Oswald." The Stovers built their home in Bradley in 1941 and continued to live in the community for sixty years. Willis Stover was also instrumental in starting the Bradley Presbyterian Church and the Bradley Lions Club.

Many families contributed to the growth of Bradley. George Barbera's father bought the old Bailey homestead. Celeste Barley, widow of Mitch Barley, said Bradley was a wonderful, caring community to live in. She and Mitch operated a store in Bradley for several years. Carl Hornsby was a coal miner and he and his wife Helen raised a family of three boys in what is now Bradley. The Hornsby's, and many other families, reared the third generation of youngsters here.

J. A. Blackburn operated a successful real estate business and reared his family in Bradley. He started his first business by driving to North Carolina on the weekends, hauling fresh produce home to sell. Later he established with partner Eddie Ferrell, a small grocery store called the "Green Lantern." One corner of the store was the first post office. Blackburn and Governor Okey L. Patteson became partners in the real estate business with offices on Route 16.

Walter Broyles built many fine houses in the community and the population began to grow. Other businesses opened up and soon the community was on the move. A portable roller skating rink would come in and pitch a tent for the youngsters to enjoy the skating. Tent revivals would sometimes fill the same field.

In December of 1948, the name was officially changed from Bailey's Cabins to Bradley. The next post office was located in a corner of Duncan's Store with Charles E. Honaker as postmaster. Honaker had just returned from World War II where he had been a prisoner of the Romanian government for four months.

The growth of Bradley spurred the need for a new, bigger post office. The new brick post office was dedicated in July of 1953. It was a real celebration for Bradley, bringing several political leaders into the community. Among these were American hero General Omar N. Bradley, for whom the community was named. With the assistance of postmaster T. E. Duncan, General Bradley posted two letters to his grandsons from the post office that bore his name.

General Bradley, accompanied by U. S. Rep. Robert C. Byrd, arrived at the Raleigh County Airport aboard an Air Force C-47 plane. Former Gov. Okey L. Patteson was master of ceremonies at the dedication program; Representative Byrd introduced General Bradley.

Bradley was welcomed to the town named for him by J. A. Blackburn, the Lions Club general chairman for the General Bradley Day program; Willis Stover, vice president of the Bradley Lions Club; and George B. Chambers, mayor of the Beckley.

The Bradley Lions Club was aided in the dedication by the cities of Mount Hope and Beckley. The day's program included a parade through Beckley's business district, the visit of the General to the Bradley post office, luncheon at the Mount Hope Lions Club, and the General's departure from the airport in the afternoon.

Bailey's Cabins: from rental cabins and a muddy toll road to a thriving community—today even Squire Jim Bailey wouldn't recognize his farm.

Bradley held a real celebration in 1953 on the opening of a new Pure Oil Company Station on Route 16, at the entrance of the Maple Fork Road. Governor Okey L. Patteson is seated in his car, Mrs. Sue Walker, Mason Walker (owner), M. B. Walker, Johnny Spade, and J. A. Blackburn. COURTESY VERNON BARLEY

Mr. and Mrs. Mason Walker in a Ford automobile with the State Police bodyguard for Governor Okey L. Patteson. COURTESY VERNON BARLEY

Gas for Governor Patteson's Cadillac with number ONE license plate. Pictured are William Lewis Hornsby and Mason Walker. COURTESY VERNON BARLEY

The men at the Pure Oil Company Station enjoyed the visit of a 1926 Model T Ford at the opening of the new station. Pictured from left to right: The men and boys to the left are unidentified, Johnny Spade at the crank, the small boy through windshield is Henry Hornsby, the first man sitting on the car is unidentified, Cecil Godby, who worked for Pure Oil Company, Sue Walker, Mason Walker, Charlie Underwood, Don Stull, and young boy unidentified. COURTESY VERNON BARLEY

J. A. Blackburn and Governor Okey L. Patteson feeding pet deer at Blackburn's home in Bradley. Many people brought their children on Sunday afternoons to view the pet deer. COURTESY VERNON BARLEY

A busy day at the grand opening of the Pure Oil Station in 1950.

The Bradley Taxi with owner Elmo Hurd leaning on his taxi and Mason Walker with Governor Patteson's car. Mr. Walker also drove a bus to Cranberry to transport miners to work. COURTESY JO ANN and M. WALKER

BUD

General View, Bud – Wyoming County: Thermo Pocahontas Coal Company.

Bud was named for Bud Adams who had a logging camp on Barker Creek and is located in Wyoming County.

CALORIC

Wyoming County: Homes owned by Smith Pocahontas Coal Company.
Caloric is now incorporated into the City of Mullens.

CARLISLE

CARLISLE MINE—TIPPLE AND PORTION OF TOWN

White Oak Fuel Co., Fayette County

Daily capacity—1250 tons coal

Carlisle mine, tipple and portion of town.

COURTESY J. W. DIXON, JR.

Carlisle – Fayette County: White Oak Fuel Company operated under the New River Company. Capacity of the Carlisle tipple was 1,250 tons daily when this picture was made in 1910.

COLCORD

Colcord depot.
COURTESY H. H. ROBINSON

COLCORD - Raleigh County: The Colcord Coal Company at Montcoal was established in 1915, but the history of the Colcord family in West Virginia goes back to 1889. That is when E. C. Colcord of Vermont, manager of Rowland Land Company, bought 70,000 acres of land for its timber. His son F. C. Colcord, a 1907 graduate of WVU, started surveying the land. Two years later E. L. Colcord, his brother, left school to help with this monumental job.

In 1911 W. H. Pettus opened the Seng Creek Coal Company at Whitesville, which served as the hub for Colcord Coal Company four years later. Pettus was named president; F. C. Colcord, general manager; E. L. Colcord, vice president and superintendent of the Stickney operations; and John Rutherford of Tennessee was vice president and superintendent of the No. 1 mine.

The Montcoal mine started running coal in August 1917 and, in time, the company became the fifth largest producer in Raleigh County. The Dorothy seam, incidentally, is named after Dorothy Rowland.

Colcord coal comes from the heart of the famous Dorothy Field of superior hard coal, low in ash and sulfur. It differs from most hard bituminous coal in that it burns more freely on the outside, consuming without cracking or burning up.

The village of Dorothy had its own churches, schools, stores, and a YMCA with splendid athletic and recreational facilities.

CRAB ORCHARD

Crab Orchard in Raleigh County operated under three companies; Crab Orchard Fuel Company, the Lilly & Hornbrook Mining Company, and later the Gulf Mining Company in 1940.

Herbert Jones and Clara Becklett, 1941.

Ralph Waddle, Dewey Blevins, and Elmer Farris with part of the Crab Orchard camp in the background. Man on the right is unknown.

CRAB ORCHARD

Crab Orchard Gulf Mining Company.
Jim Fitzpatrick and Homer Persinger on the "Dinky" locomotive, whose last run was in July 1952.

Clayton Cooper, Homer Persinger and Alfred Persinger.

Joe L. Smith, picking bone.

The "Dinky."

CRANBERRY

CRANBERRY MINE — TIPPLE AND POWER HOUSE, DURING CONSTRUCTION

Cranberry Fuel Co., Raleigh County

Daily capacity — 1250 tons coal

Cranberry – Raleigh County: Mine of the Cranberry Fuel Company.
Tipple and powerhouse during construction.
COURTESY J. W. DIXON, JR.

DOROTHY

Dorothy – Raleigh County: The Big Supply Company store 1914-15. The small building below the company store is the band room.

Cecil Phelps in front of the John Daskolas store, 1924. Mr. Phelps is the father of Reva Mae Robinson of Dorothy. Also see the photo in Sarita, as the store straddled the line between the two towns.

PHOTOGRAPHS OF DOROTHY COURTESY HORACE H. ROBINSON.

Dorothy grade school 1924.

DOROTHY

Delivery wagon of the C & O Fuel Mine Operations, 1916-17.

Dorothy was a mining village on the Clear Fork of Coal River, established by the Four States Coal & Coke Company. In 1916 it had three stores, one hotel, and a Mission School, established by the Presbyterian Church, located one mile east of Dorothy.

Fire truck of the C & O Fuel Mine Operations, H. B. Husband standing beside the truck in1930.

Dr. Benjamin Lewis Pettry, called "Dr. Ben," covered Dorothy, Sarita, and part of the time, Ameagle. "Check off" for the doctor was $1.00 a month. If he delivered a baby, the charge was $10.00. Until he purchased a car, he rode a bicycle to make his calls.

Dr. Ben sent this photo to his bride-to-be in Georgia, and wrote, "I've put a cross over our house. It is exactly like the one nearest the store – 6 rooms, plastered." They were married in 1923.

DOROTHY

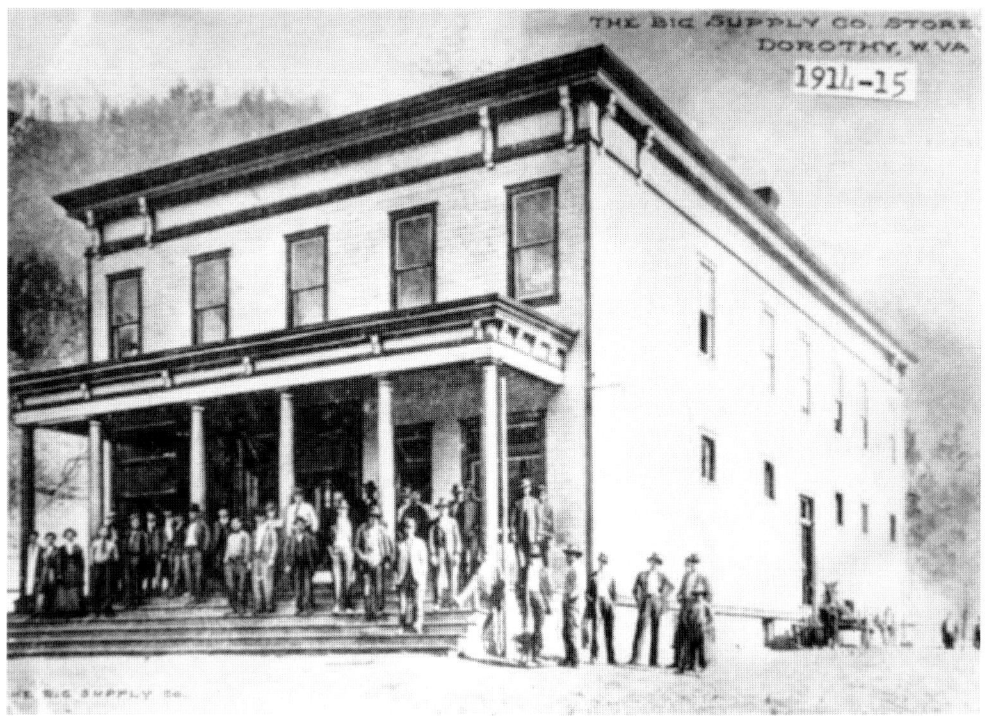

The Big Supply Company store 1914-15.

Dorothy company store, bottom tipple, and restaurant.

Leona Gillispie's 4th grade class, Dorothy grade school (1932).

Clarence Nutter, driver for Wilhoit Dry Cleaners (1928).

Zachari Karantonis had the first store in Dorothy. Pictured are, Cost Devalkis, Mrs. Zachari Karantonis and baby John, Georgia Karantonis, Stella Zacharias, Mr. Zachari Karantonis (with tie), Cecil Taylor, two children are unknown but child with dark hair is Billy Karantonis.

Interior of Shibley, Shibley & Kannon Store.

Lady with broom and man beside her are Mr. and Mrs. Samuel Robinson, father of Horace Robinson.

1910 Dorothy band.

First graduation class of Clear Fork High School – 1929. Ernest J. Wauhop, Principal.

Eldest girl is Helen Kincaid, with her brother and sister.

Ladies first aid team of the C & O Fuel Mine Operations in 1932. Edna Robinson, Josephine Harris, Icia Kirk, Nellie Cook, Faye Peters, and Rosie Pizzuto.

Bruce "Ted" Cline, Horace Robinson and dog Jane, 1932, Kayford, WV.

Players from Ameagle making up the Dorothy team in the mid 1940s. Bottom row: Harvey Stover, Hermit Stover, Eugene Massey, James Kanasky, and Buster Stanley. Middle: Ben Henry and Arthur Short. Top row: John Lapri, Quinton Scarbro, Orville Dillon, Basil Legg, Shelton Stover, Lonnie Evans, Jay Miller, Ishamel Frailey, and John Heller.

Dorothy Fire Department: Bill Kavinsky, Lewis Matheny, Cecil Taylor, John Kirk, Lewis Harvey, Ash Cook, and Clyde Markham, who was the Special Agent for Criminal and Civil Matters.

Men receiving their last pay check when the mines closed down in 1951.
Trauax Tracer Coal Company, Shamrock 33 Mine.

The second store of Zachari Karantonis, early 1940, operated by his son, Manuel.

A Sunday afternoon poker game.

EAST GULF

East Gulf – Raleigh County: Store and office building. C. H. Mead Coal Company organized in 1920, advertised as "Laundered Coal," and in the first six months of operation, produced 34,665 tons. They shipped on the C & O and the Virginian Railways. The operating offices of the C. H. Mead Coal Company were located in Beckley. Below is a sectional view of East Gulf.

Mines Numbers 3 and 4 of New River Collieries Company at Eccles.

Eccles – Raleigh County: Located on Millers Camp Branch of the Coal River, on the Virginian Railway, and Piney Branch of the C & O Railway. On January 1, 1915, its postmaster reported a population of 1,500 people. On April 28, 1914, an explosion in No. 5 mine killed 182 men.

Mines Numbers 1 and 2 of New River Collieries Company.

ECCLES

View of Eccles.

Mine No. 6 of New River Collieries Company.

EDWIGHT and EUNICE

Town of Edwight – Raleigh County.

Princess Dorothy Coal Company in Eunice was well known for the high quality of its coal mined from the Whitesville seam and called Robin Hood coal. It was popular for steam and domestic purposes. Smokeless Fuel Company sold Princess Dorothy coal to markets in Atlanta, New England, and the Great Lakes. COURTESY HORACE ROBINSON

Eunice – Raleigh County: Eunice Fire Department crew of the C & O Fuel Mines Operations in 1935. Shown from the passenger side, George Baxter, and at wheel, Elvin Miller, standing, Charlie Maye, Paul Matos, and Orville Miller.

FIRECO

Fireco – Raleigh County: Leckie Fire Creek Coal Company.

Miners' homes, Lillybrook No. 2 at Fireco.
Fireco, a name indicating heat, was thought to have advertising value.

Store and office building.
Leckie Fire Creek & Douglas Coal Company.

Lillybrook #2 store and office.

GARY

Gary – McDowell County: United States Coal & Coke Company.
The largest tipple in the world on the day before it was razed in 1991.
COURTESY E. F. REYNOLDS COLLECTION

U. S. Coal & Coke tipple – ten minutes before final demolition in 1991.
E. F. REYNOLDS COLLECTION

GLEN JEAN

Glen Jean – Fayette County: In 1890 Justus Collins leased land for coal mining purposes and soon founded Collinsville. In 1897 the name was changed to Glen Jean. The above store of the Collins Colliery Company was erected in 1893. C. W. Hardy was serving as store manager when this building burned on December 17, 1900. COURTESY FRED FRISK

STORE BUILDING AND SUPERINTENDENT'S HOUSE
Collins Colliery Company

Merchandise $40,000
Sales per year $114,000

Collins Colliery Company - Store building and superintendent's house.

Four months after fire destroyed the previous store, this new look-alike building opened for business. A busy store doing $144,000 sales per year, this building also fell victim to fire several years later.

Justus Collins' mines were located on Dun Loup Creek, a tributary of the New River and on the Loup Creek Branch of the C & O Railway, about seven miles from Thurmond. Red Ash coal in the Sewell Seam was mined. Justus Collins served as general manager, George Daniels, superintendent, and Jeff Saunders, mine boss.

Thomas Gaylord McKell came to Fayette County to investigate his wife's interest in a tract of 12,500 acres of land her father gave as a wedding present. After investigation of the property, he proceeded to purchase the interest of the other heirs in the 12,500-acre tract, as well as many adjacent properties, until he rounded out a solid boundary of approximately 25,500 acres under laid by Sewell and Fire Creek seams. He had extensive surveys made of the land and made plans for a railroad to go from Thurmond to MacDonald. He then secured, by contract with the C & O Railway, the construction of the Dun Loup Creek branch for the purpose of developing the Loup Creek coal field. This branch railroad was completed in 1893, and the first car of coal was shipped from the mine of the Collins Colliery Company (a McKell lease) at Glen Jean on November 7, 1893.

COURTESY J. W. DIXON, JR.

Blacksmith Shop in Glen Jean.
The Bank of Glen Jean was built in 1909 on this same property.
COURTESY FRED FRISK

In 1900, Mr. McKell decided to develop some of his properties himself and organized the McKell Coal and Coke Company for this purpose. During the period from 1900 to 1905, he had the following mines opened: Derryhale, Kilsyth, Oswald, Graham, and Tamroy.

GLEN JEAN

In 1896, Mr. McKell built the Opera House at Glen Jean from the plans of architect, Frank L. Packard of Columbus, Ohio. In 1901 he built the Dun Glen Hotel and viaduct at Thurmond. Mr. Thomas G. McKell died in Atlantic City, NJ, in 1904.

Mr. William McKell took up his residence at Glen Jean in 1893 and after the death of his father in 1904, took general charge of the property and the various enterprises connected therewith. He organized the Bank of Glen Jean in 1909. COURTESY FRED FRISK

GLEN JEAN

The Kanawha, Glen Jean and Eastern Railroad, a McKell enterprise, which had been built from Tamroy to Glen Jean was extended in 1910 another six miles, from Mount Hope to Pax. This made it possible for a great many of the mines on Dun Loup Creek to ship their products over the Virginian Railway. COURTESY FRED FRISK

The streetcar of the Kanawha, Glen Jean and Eastern that carried passengers over their rails from Tamroy to Mount Hope or Thurmond. COURTESY FRED FRISK

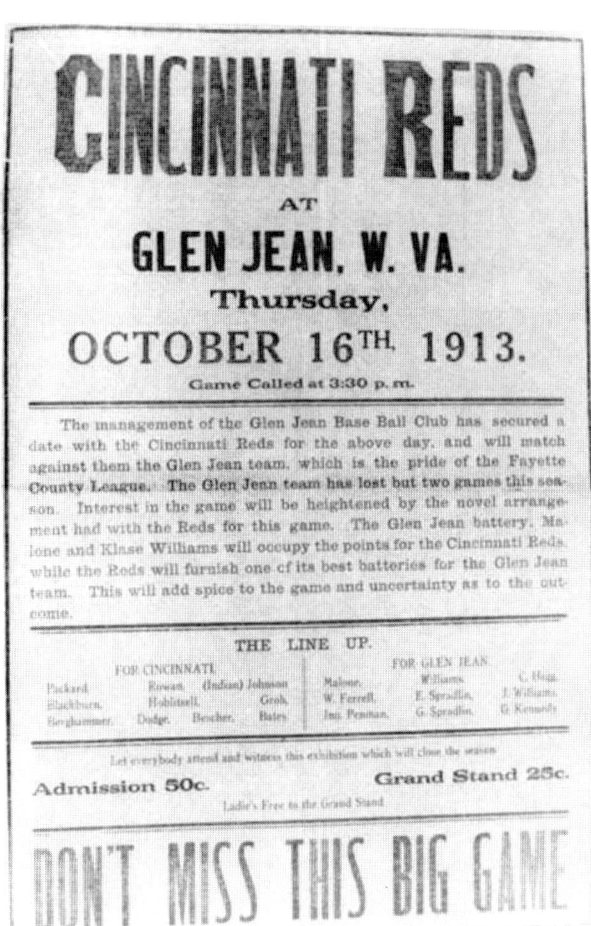

Period advertisements.

Collins Colliery Company – tipple and coke ovens.

Klondyke Mine of the Dunn Loop And Coke Company.

Glen Jean Peace Keepers.
Circa 1905. COURTESY W. O. TRAVEY PRINT – E. F. REYNOLDS COLLECTION

William McKell owned nearly all the houses of Glen Jean, which covered the level land along the creek and extended far up against the slopes of the mountains. The town had churches, schools, a bank, a well-stocked department store, two coal company stores, a drug store, two garages, a post office, photography gallery, dentist's office, barber shop, and a gas station.

With the Kanawha, Glen Jean and Eastern Railway, Glen Jean became the hub for shopping and services for the miners' families.

GLEN ROGERS

Mines of Raleigh Wyoming Coal Company.

Glen Rogers – Wyoming County: On May 31, 1918, Raleigh Wyoming Coal Company obtained a 40-year lease on about 9,108 acres lying partly in Raleigh and partly in Slab Fork district of Wyoming County. The work of sinking the shaft started in the summer of 1920. This shaft was 650 feet deep; the site was swampy and the rock was hard. The roads became muddy and impassable during the winter season. It would take two years and $2,000,000 to complete construction of shafts, mining camp, office, and other necessary buildings. Brick for building purposes was made on the grounds. Production started in 1922.

GLEN ROGERS

The Raleigh Wyoming Coal Company was under the leadership of H. H. Rogers and G. W. Hyman, a Boston financier. It was on Mr. Hyman's recommendation that the Virginian coal carrying cars were built to hold 110 tons instead of the customary 55-ton capacity of other railroads. Glen Rogers became the largest coal operation on the Virginian line. Production of coal started in 1922. In 1960 the company was adjudged bankrupt and its assets sold.

Original log cabin located on property of Raleigh Wyoming Coal Company.

Superintendent's Home of the Raleigh Wyoming Coal Company.

Snow on May 10, 1923.

GLEN WHITE

Company store of E. E. White Coal Company.

Glen White – Raleigh County: Mr. E. E. White, born in England, came to West Virginia and obtained a lease from Beaver Coal Company. In 1904 he opened the Beckley mine near the town of Beckley. In 1906 he sold this mine to the New River Coal Company and obtained another lease on which he opened Glen White and Stotesbury Mines in 1909 and 1910. In 1925 he sold the mines and retired to Pennsylvania.

Glen White, a very progressive community, with neat fences and real sidewalks. Glen White was named for the president of the coal company, E. E. White.

GLEN WHITE

Bible class and Sunday school.
E. E. White Coal Company, Glen White, West Virginia.

Recreation building of the E. E. White Coal Company.

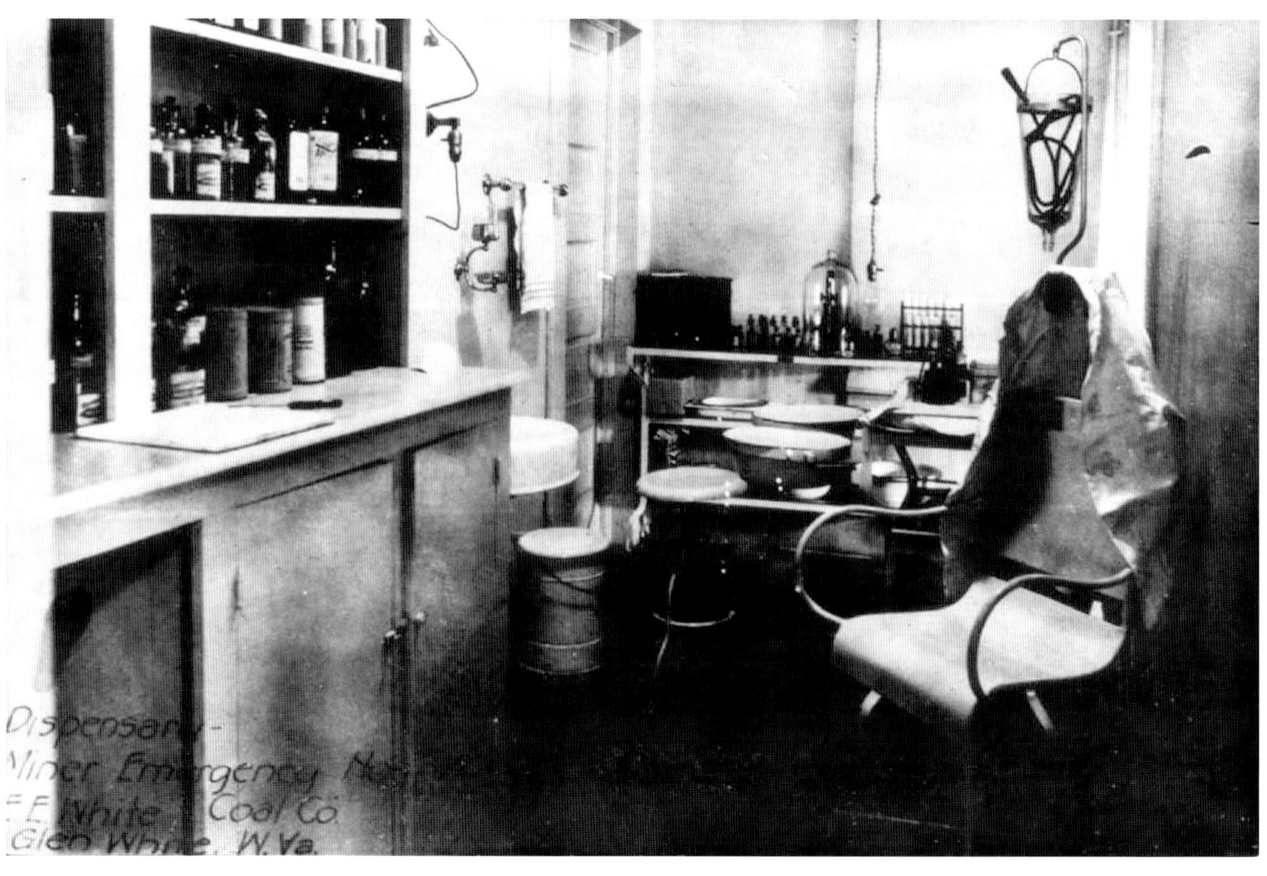

Dispensary-mine emergency office.

GLEN WHITE

General view of tipple and powerhouse.
E. E. White Coal Company

HARVEY

HARVEY OPERATION—TIPPLE AND COKE OVENS

Harvey Coal and Coke Co., Fayette County

Daily capacity — 1000 tons coal
125 tons coke

Harvey tipple and coke ovens.
Harvey Coal & Coke Company.
COURTESY J. W. DIXON, JR.

Harvey – Fayette County: The Harvey Coal & Coke Company began operations in 1893 and was a subsidiary of the ever growing New River Company under the leadership of Samuel Dixon. Their first post office was Bissel. The Harvey mine was located on Dunn Loup Creek, just five miles from Thurmond. M. M. Warren was the first superintendent, with A. P. Gibson as the mine boss. The general manager of the Harvey operations was J. B. McGuffin.

The Harvey mines were drift mines located on the Loup Creek branch of the C & O Railway. Red Ash coal was mined in the seam known as the Sewell, which varied from 42 to 66 inches.

Harvey Coal & Coke Company closed the mines in 1945 after mining 5,624,084 tons of coal that year. The company store closed in 1954.

Store and office building – East Gulf Coal Company.

General view of Helen – Raleigh County.

East Gulf Coal Company – Miners' homes.
Helen was named after the daughter of G. W. Stevens, president of C & O Railway.

East Gulf Coal Company club house.

ITMANN

General view of Itmann – Wyoming County.

In 1918 Pocahontas Fuel Company opened a drift mouth mine and this camp was named Itmann. It was a sizeable community complete with about one hundred houses for officials and workmen, separate schools for white and black children, a company store, post office, and church.

Itmann was named for Isaac T. Mann, an early coal operator and owner of Itmann Coal Company. Itmann flourished for several years but became inactive and was partly dismantled about 1958.

View of Itmann.

ITMANN

Itmann mine processing plant.
COURTESY E. F. REYNOLDS COLLECTION

JONBEN

Wilton Smokeless Coal Company – Jonben school house – Raleigh County.

Store and office building – Wilton Smokeless Coal Company.

KILSYTH

Kilsyth - Fayette County: McKell Coal Company's mine operations. Shown above is part of the town, coal tipple, powerhouse and company store. At the time of this picture, about 1910, the company worked 350 to 400 men. Daily output was 1,500 tons. Kilsyth is located on the Kanawha, Glen Jean & Eastern Railroad, connecting with Virginian Railway at Pax.

Kilsyth was a drift mine located near the town of Mount Hope on Dun Loup Creek about 10 miles from Thurmond on Loup Creek Branch of the C & O Railway.

Red Ash coal was mined from the Sewell seam. Thomas Nichol was superintendent and Thomas A. Lewis, Jr. was mine boss.

KILSYTH

LAYLAND

Layland – Fayette County, 1919.
COURTESY H. C. WELLS

About the year 1891, William Beury, Cooper and Company bought a large section of land, and preparations began for the opening of mines. The post office was established and called Gentry in honor of the first settler, but when Beury, Cooper and Company later sold out to the New River & Pocahontas Consolidated Coal Company, the name was changed to Layland. This company immediately began the erection of modern frame buildings as homes for its men. The store building was of cut stone and said to be the finest in the state.

In 1926 Layland was reported as a very desirable place to live, with comfortable, pleasant dwellings, a magnificent superintendent's home, a large concrete tennis court, schools and two churches.

LAYLAND

View of Layland – 1919.
COURTESY H. C. WELLS

On March 2, 1915 a tragic explosion at the Layland mine killed 114 men inside and one man outside. The force of the explosion from the drift mouth shook buildings and broke windows in the vicinity. Some of the 114 miners were killed by the blast, others by slate falls, and forty-four died from suffocation.

The store porter passing the drift mouth at a distance of 100 feet at the time of the explosion was hurled against a post and killed.

Fifty-four men escaped alive from the mine. Seven came out from two to five hours after the explosion, five more escaped unassisted on March 6 and forty-two others were rescued the same day.

The explosion extended over a major portion of the mine and the rescue and recovery efforts were difficult and dangerous.

LAYLAND

Layland – New River & Pocahontas Consolidated Coal Company.
COURTESY H. C. WELLS

LAYLAND

Layland – New River & Pocahontas Consolidated Coal Company.

The area's first settler, H. H. Gentry brought his family from Albemarle County, Virginia. He selected a high plateau where he built a log house surrounded by a split rail fence. He cleared some of the land and raised crops of corn and pumpkins. For the first few years they lived principally on the wild game that was plentiful in that area.

Two years later, Mr. Gentry's brother-in-law, Ira Gibson, moved into the area with his family. Gentry and Gibson brought all their belongings in covered wagons and on packhorses over the mountain, and across the valleys and rivers, driving their cattle along with them. Hard working men, they soon had fields of corn and wheat. Herds of sheep and cattle roamed the hills and were guarded from wolves and bears by the men and boys of the families.

With the coming of the coal industry, the fields soon gave way to mines, and houses covered up the slopes of the mountains.

LEGO

Miners' homes at Lego – Fire Creek Smokeless Fuel Company.

Lego – Raleigh County: Store and office.

Superintendent's home.
Fire Creek Smokeless Fuel Company & Laurel Smokeless Coal Company

Miners' homes.

LILLYBROOK

Miners' homes at Lillybrook – Raleigh County.

Lillybrook Coal Company was owned by Prince Lilly and John Hornbrook, who founded the company and gave the community its name.

Tipple of the Lillybrook Coal Company.

Lillybrook Coal Company - Store and office.

Sectional view of Lillybrook.

LIVELY

Water Tank at Lively – Fayette County.

LOCHGELLY

Lochgelly – Fayette County: Samuel Dixon named this town Stuart for a relative. After the great mine explosion of January 29, 1907, succeeding management changed the name to Lochgelly in an effort to help people forget the mine disaster. The last mine at Lochgelly was closed by the New River Company in 1960.

See also pages on Stuart.

LYNWYN

Lynwin - Raleigh County: Lynwin Coal Company opened for business in 1906 in Mistletoe and comprised some 600 acres. The Beckley seam was mined with the plant located near the Winding Gulf post office. According to the August 30, 1917 publication of the *Register's Industrial Review*, the Lynwin community consisted of 32 three-room houses, 24 four-room houses, and 7 five-room houses.

The capacity of the plant was 500 tons of coal daily. The powerhouse was of frame construction and produced 412 horsepower. The tipple was constructed of timber with picking tables and loading boom equipment, and the mines were entered by drift mouth.

In 1917 the company operated a motion picture theatre for the entertainment of its employees. Employees numbered 125 men.

The Virginian Railway reached Lynwin around 1911. Lynwin Coal Company sold its holdings to Winding Gulf Collieries about 1930.

LYNWIN

Lynwin company store.

On the Lynwin Coal Company store steps in 1931 or 1932. Front row: left to right, first man is unknown, Charles Gwinn, store manager, Robert Hurt, Mr. Jarrell, Bill Peters, and Jess Moore. Back row: Brown Thompson, Andy Meickle, Clarence Meadows, Carl Moore, Ernest Palmer, and Leonard Carr.

Lynwin school children in1931. Mae Walker, the teacher, is at far right.

MacALPIN

John Laing was born in Lanarkshire, Scotland. When he arrived in Raleigh County he leased 1700 acres and formed his company and community in 1906. He began operations with a capital of $100,000. He named his community MacAlpin, and his company, MacAlpin Coal Company after his mother, whose maiden name was MacAlpin.

In 1920 MacAlpin mined 204,174 tons of coal and employed 109 men using 12 mining machines. They also had 14 pick miners who produced another 38,664 tons. The mine worked the Beckley seam of coal. The superintendent was A. W. Laing and W. F. Riley was mine foreman.

School house at MacAlpin.

Cars and garages of the employees.

MacAlpin auditorium.

Miners' homes.

Miners' homes.

MacAlpin Coal Company on the Winding Gulf Branch of the Virginian Railway, showing the steel tipple, and powerhouse at left. COURTESY E. F. REYNOLDS COLLECTION

Residence in MacAlpin.

MacALPIN

A very early picture of MacAlpin.

Bailey Wood Coal Company office at MacAlpin.

View of Bailey Wood Coal Company.

MacDONALD

General store of the MacDonald Colliery Company.
COURTESY J. W. DIXON, JR.

MacDonald – Fayette County: MacDonald Colliery Company operated a drift mine on Dun Loup Creek about eight miles from Thurmond. Red Ash coal of the Sewell seam was mined. Samuel Dixon was general manager and Wayman Hart was mine boss.

The New River Company's offices were located in MacDonald. MacDonald is now incorporated into the city of Mount Hope.

In the young days of organizing, safety first was always stressed with the New River Company, who encouraged their miners and office personnel with safety programs. A festive day with friends, food, and fun was set aside to show their safety programs. The first aid teams from all mines would assemble for an aid recovery test. Prizes were awarded to the winners. The New River Company was the first in the entire United States to take up a full and complete course in first aid and mine rescue work. The *History of Fayette County, WV*, published in 1926, reported that the first aid team won the national championship at Denver, Colorado, and the mine rescue team won the national championship at St. Louis, Missouri.

Music was encouraged through the formation of the New River Band. The New River Company was confident that music offered a solution to the pressing problems of their day, and would prove helpful in the life of tomorrow. In 1930, under the direction of Arthur Cook, the New River Band consisted of thirteen students. A picture of the New River Band two years later showed Mr. Cook and a group of twenty-seven youngsters with both boys and girls.

Residents of all the New River Company's towns were encouraged to beautify their yards by planting gardens and painting fences. The company planted trees in each yard and encouraged the residents to care and protect the trees to enhance the community.

Boy Scout troops were formed. The company published *New River Company Employees Magazine*" to cover the news from all of the New River Company communities.

SCHOOL HOUSE AND PRESBYTERIAN CHURCH
Macdonald Colliery Company, Macdonald, West Virginia

School house and Presbyterian Church.
COURTESY J. W. DIXON, JR.

MacDONALD

COKE OVENS — MACDONALD COLLIERY CO.

Macdonald, West Virginia — Daily capacity—150 tons coke

Coke Ovens – MacDonald Colliery Company.

MACDONALD MINE—TIPPLE

Macdonald Colliery Co., Fayette County — Daily capacity—750 tons coal, 150 tons coke

MacDonald mine tipple.

MacDONALD

Town of MacDonald, established in 1893, showing tipple on the hillside.
COURTESY J. W. DIXON, JR.

MacDONALD

The original MacDonald Colliery Company store served as a part of the main office of the New River Company. This store was established in 1893 and burned March 23, 1925.

The New River Company, general office staff, July 14, 1912, MacDonald, WV.

Store and office of the Micajah Pocahontas Coal Company – Wyoming County.

Tipple of Pocahontas Coal Company.
Micajah was named for an early settler, Micajah Bailey.

Club house.

Virginian coal train on Bridge No. 68.

MONTECARLO

General view of Montecarlo – Wyoming County: Monticello Smokeless Coal Company.

Miners' homes and school building.

MORRIS CREEK

Tipple of Mecca Mines on Morris Creek near Montgomery – Fayette County.
COURTESY ROBERT P. BODNAR

MOUNT HOPE

Town of Mount Hope – Fayette County.
COURTESY J. W. DIXON, JR.

The first family to settle in what is now Mount Hope came from Scotland in 1805. As late as 1870 only three families comprised its entire population.

In September 1893, a connecting link of the C & O Railway was made from Thurmond to Mount Hope. The mining developments began with the advent of the railroad and after that date the population increased and mines opened in rapid succession. The town was incorporated in 1895 and granted a City Charter by the State Legislature in 1921.

Mount Hope had grown to be a prosperous village of some 12,000 people; with all signs indicating continued rapid growth. On the morning of March 24, 1910, the news was flashed over the wires throughout the county and state, "Mount Hope is burning up." About forty business buildings, and 150 dwelling houses were consumed. The town was left a mass of blackened ruins. No sooner had the embers cooled than the debris was cleared away to make way for new and better buildings. Many of the businessmen erected temporary quarters and were ready for business in a few days. Just a year and a half had elapsed since the town was in ruins, and in that time almost every business that burned had been replaced by a larger structure of a type that "defies flames," reported *The Fayette Journal*, November 2, 1911.

Mount Hope was the hometown of the twenty-third Governor of the state, Okey Leonidas Patteson. Born at Dingess, a coal town in Mingo County, his family moved to Mount Hope when he was a one-year old. His father opened a general merchandise store and later entered the hardware business.

The fire that swept Mount Hope resulted in the loss of both Mr. Patteson's store and home. The National Guard moved into the charred ruins of Mount Hope to issue emergency supplies and tents for dwelling. One of the tents was issued to the Patteson family of four.

Patteson did rebuild his store and a temporary home. Ultimately, he build a beautiful permanent residence in which his well-known son later made his home.

The New River Company was the largest holding company operating in the New River field. New River was a West Virginia corporation, promoted and organized in 1906. Samuel Dixon, who was the promoter of this great aggregation of industrial capital, serving as president and active head of the empire for many years. In addition to his interest in the New River Company, Samuel Dixon held stock in the Dixon-Pocahontas Coal Company, the Gulf Smokeless Coal & Coke Company, Prudence Coal Company, and Sugar Creek Coal Company.

Mount Hope is located in the southern edge of Fayette County, just west of the junction of Sugar and Dun Loup creeks, on the Kanawha, Glen Jean & Eastern Railway, and the Loup Creek Branch of the C & O Railway.

Office building for the New River Company, MacDonald now incorporated into Mount Hope.

MOUNT HOPE

A young Mount Hope, rebuilding after the great fire of 1910.

Mount Hope Bakery on Mills Street delivered bread from Mount Hope to Scarbro, Whipple, Oak Hill, Hill Top, and Glen Jean.

MOUNT HOPE

Sugar Creek Coal & Coke Company, Mount Hope.

Sugar Creek began shipping coal in 1896. Later the New River Company purchased the mine. According to the 1903 *Department of Mines Annual Report*, Sugar Creek was a slope mine located on the Big Loup Creek branch of the C & O Railway, about ten miles from Thurmond. Average number of employees, 100-125 men and boys. Pick miners who received 45 cents per ton mined all of the coal.

Dartmont Coal Company, Mount Hope.

OAKWOOD

OAKWOOD MINE—TIPPLE AND POWER HOUSE

White Oak Fuel Co., Fayette County — Daily capacity—1000 tons coal

Oakwood mine, tipple, and powerhouse of the White Oak Fuel Company.

OAKWOOD MINE TIPPLE

White Oak Fuel Co., Fayette County — Daily capacity 1000 tons coal

Oakwood mine & tipple – Fayette County.

The Oakwood mine had a daily capacity of 1,000 tons in 1911. Oakwood ceased operations in 1960.

PARRAL

PARRAL MINE

Stuart Colliery Co.

Daily capacity—1200 tons coal

Parral mine tipple of the Stuart Colliery Company – Fayette County.
COURTESY J. W. DIXON, JR.

The name Parral town was changed to Summerlee on February 28, 1912. (See also pages on Lochgelly and Stuart.)

Long Branch Coal Company near Pax – Fayette County: Mined 67,710 tons of coal in 1913.

First Long Branch tipple 1912-23.

PAX

Theatre at Pax.

Mr. Jeff Tyree, Bank of Pax 1918.

FREIGHT AND PASSENGER TRAIN
Piney River and Paint Creek Railroad Company

Freight and passenger train.
COURTESY J. W. DIXON, JR.

Store and office – Ragland Coal Company.

General view of Pemberton – Raleigh County.

PHILLIPS

Phillips – Raleigh County: Miners' homes – Pembertson Coal & Coke Company.

Powerhouse 13-ton main line motor and engine.

Miners' homes – Pickshin Coal Company – Raleigh County.

Company store and office.

PRICE HILL

Price Hill – Fayette County: mine, tipple, and powerhouse. Price Hill Fuel Company.
COURTESY J. W. DIXON, JR.

Bridge construction – Price Hill Branch Of White Oak Railway.
The bridge spans Sugar Creek and the tracks of the C & O Railway.

THE NEW RIVER CO.

No. 33 .... J F Elkins .... 193....

In Account With

THE NEW RIVER CO., NO. 14
PRICE HILL, W. VA.

Pay Roll Clerk Stamp Name of Company in Space Above

JUL 31 1939

For month of .... 193....

Cr.			Dr.		
By........Cwt. @			To Store	8	
By........Cwt. @			To Overdrawn Last Month		
By........Cwt. @			To Cash		
By........Cwt. @			To Rent	7	50
By........Hrs. @			To Electric Lights	4	30
By........Hrs. @			To Water		
By........Hrs. @			To Coal		
By........Track Footage			To Hauling		
By........Yards @			To Safety Lamp Charge		
By........Yards @			To Smithing		
By........Yards @			To Hospital		
By........Yards @			To Relief Fund		
By........Board or Labor			To Insurance		
By........Cash			To Doctor		85
Skelton	59	04	To Union		
			To Fed. O. A. B. Price Hill / C	5	00
			Macdonald	15	00
			To Board or Labor		
Total			Total	40	65
To Balance Due Company			By Balance Due Workman	18	39

Form 35 500 Pds. 12-37 (37) 8521 E. & O. E. Liable for any Subsequent Indebtedness

Retain This Statement

A payroll statement dated July 1939 for Mr. J. F. Elkins, grandfather of Lois Moore of Beckley. Although this statement dates from much later than the pictures in this book, it is an interesting document of a past era.

Samuel Dixon's empire of the New River Company reached in many directions. Price Hill was the location he chose to build his home.

PRINCE

Prince railroad station – Fayette County.

Tracks at Prince.

PRINCE

The Chesapeake & Ohio Railway Bridge, which spans the New River at Prince.

PROSPERITY

Prosperity – Raleigh County: Mine, tipple, and powerhouse during construction.
Cranberry Fuel Company.
COURTESY J. W. DIXON, JR.

In 1906 Samuel Dixon of the New River Company bought the Beckley Coal & Coke Company and the Mabscott Coal & Coke Company. Land was then purchased that was to become the communities and mines of Sprague, Skelton, Cranberry, and Prosperity.

In order to reach the future mine locations, Dixon built the Piney Branch of the C & O Railway, about seventeen miles from the Prince Station.

PRUDENCE

Prudence mine – Prudence Coal Company.
COURTESY J. W. DIXON, JR.

Prudence – Fayette County: This was a drift mine located on Dun Loup Creek, a tributary of the New River, and was on the Loup Creek Branch of the C & O Railway. The coal mined at Prudence was Red Ash and the seam worked was the Sewell, which varied in thickness from 52 to 66 inches.

PRUDENCE

1914 photo of operations at Prudence Coal Company.
Prudence was named for Prudence McGuffin, daughter of operator John A. McGuffin.

Quinnimont – Fayette County: The Quinnimont Hotel opened for business in April, 1908. It was reported to be one of the best hotels in the state. The proprietor was W. R. Teppett. COURTESY FRED FRISK

QUINNIMONT

Quinnimont Community – circa 1929.

Jacob Smith was the first settler on what is the present site of Quinnimont in 1827. After the building of Stretcher's Neck Tunnel by the railroads, coal mines were opened. In 1870 the Quinnimont Charter Oak & Iron Company was organized for the purpose of building and operating an iron furnace. The furnace manufactured pig iron and operated for three years. A boarding house was established for the iron furnace employees.

Joseph L. Beury established the town of Quinnimont in 1873. He shipped the first coal out on the C & O Railway in September 1873.

Quinnimont, being located near the junction of the Piney River Railway and the Laurel Creek Branch Road, soon became a railroad center. It was a prosperous community with a store, post office, church and two schools. It was very accessible, as it had daily two local trains each way over the C & O to Beckley and Lester, and one daily up Laurel Creek to Layland.

"Fanny" and the Quinnimont station.

This passenger train, affectionately called "Fanny," is backing out of the front track getting ready to go west en route to Beckley. "Fanny" began operating from Prince and through Piney Creek Gorge on the C & O line in 1903. Service was discontinued to Cranberry in 1939. The Beckley-Prince service began in 1903 and was discontinued in 1949.

RALCO

Ralco store and office of the Raleigh Fire Creek Coal Company.

General view of Ralco – Raleigh County.

RALEIGH

Near the Raleigh Station on Upper Piney Creek – Raleigh County.

Raleigh Coal & Coke Company was formed in 1898 under the leadership of Logan M. Bullitt, of Philadelphia.

The first buildings erected were a store and a large boarding house. Slowly, a community grew up around the swampy bottom that had been cleared by a lumber company.

The company's greatest growth was during the era of Colonel Ernest Chilson, vice president from 1906 until his death in 1931.

RHODELL

Store and office building – Rhodell Coal Company.
Rhodell – Raleigh County: Was founded in 1907 and named for one of the founders, I. J. Rhodes. An earlier name was Rhodesdale.

Rhodell school house.

Sarita – Raleigh County: Sarita Mine Bottom Tipple crew 1925 – C & O Fuel Mine Operations.

Man on ladder, unknown. Standing on car is Albert Pauley; on steps are Ezra Short and Nick Decilias and in car, Ted Cline. Man on horse is Harry Kirk. COURTESY HORACE ROBINSON

Sarita bottom tipple.

SARITA

Sarita – circa 1925.

In front of the John Daskolas Store. The man standing in the middle is Zachari Karantonis. The store burned in 1930. Also see the photo in Dorothy, as the store straddled the line between the two towns. COURTESY HORACE ROBINSON

Silent movie theatre at Sarita in 1921. The Model T Ford has posters of coming attractions. Shown in the doorway is the operator of the theatre, Butch Caulfett; Annis Phelps, reel operator; and Homer Watts, Leige Short, and Raymond Phelps. COURTESY HORACE ROBINSON

SCARBRO

GENERAL STORE — OFFICIALS' HOUSES IN THE BACKGROUND
White Oak Fuel Company, Scarboro
The aggregate of the sales in the nineteen stores of The New River Company, of which the above is a type, exceeded in the year 1906, $900,000. The stock in each store was turned over on the average 4½ times in the course of the year.

Scarbro – Fayette County: General store in White Oak Fuel Company.
COURTESY J. W. DIXON, JR.

Samuel Dixon named Scarbro for a town in Yorkshire, England. The store, built around 1905, carried one of the most complete stocks of general merchandise to be found in the area. The first company store for Scarbro was opened in a small building in 1899, and was replaced by the one pictured above. The first manager of the new store was Lem Douglas.

A man very much respected for his fair and honest marketing, W. C. Bishop was made manager in February 1919. He remained manager until 1953 when he was promoted to manage all of the New River Company's stores. There were sixteen commissaries at that time.

Scarbro had an amusement hall that housed several businesses to serve the community; a theatre, barber shop, pool room, and drug store. The company store and post office were located nearby.

SCARBRO

Town of Scarbro- incorporated in 1904.
PHOTOGRAPHS OF SCARBRO COURTESY J. W. DIXON, JR.

The development of the mines caused Scarbro to spring up quickly. Houses of worship were erected by the Baptists, Methodists, and Catholics. By the early 1920s, Scarbro was replete with general stores, garages, filling stations, lunch counters, barbershops, a train depot, and a post office.

In 1899, Samuel Dixon built the White Oak Railway, which at first extended from Glen Jean to Carlisle. This extension opened up mines at Wingrove, Scarbro, Whipple, Oakwood, and Carlisle. A further extension of the railroad to Stuart allowed the mines at Summerlee (Parral) and Lochgelly (Stuart) to ship coal out of their mines.

SCARBRO

SCARBORO MINE—TIPPLE AND POWER HOUSE

White Oak Fuel Co., Fayette County — Daily capacity—1500 tons coal

Scarbro mine, tipple, and powerhouse – White Oak Fuel Company.

ELECTRIC LOCOMOTIVE USED IN HAULING MINE CARS

Electric locomotive used in hauling mine cars.

Joint station of White Oak Railway & Company and C & O Railway at Scarbro operations of the New River Company.

Passenger train, White Oak Railway Company.

SEWELL

Sewell – Fayette County: Sewell Lumber Company filed bankruptcy in early 1914 and was purchased by Babcock Coal & Coke Company. The lumber business flourished for fifteen years. Logging crews lived in company camps composed largely of shanty cars. A company cook prepared huge quantities of food for the laborers. Camps were moved frequently to keep pace with the advancing onslaught against the virgin timber. In fact, during the 20-year operating period, as many as 40 camps were used on the 40,000 acres that Babcock Coal and Coke Company logged.

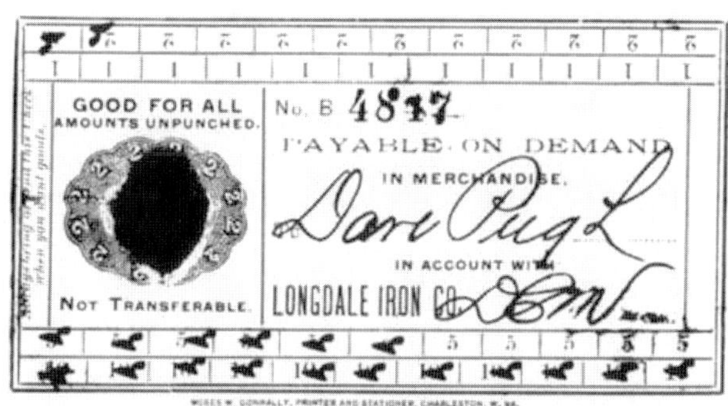

GOOD FOR ALL AMOUNTS UNPUNCHED.

No. B 4847

PAYABLE ON DEMAND IN MERCHANDISE.

Dave Pugh

IN ACCOUNT WITH LONGDALE IRON CO.

NOT TRANSFERABLE.

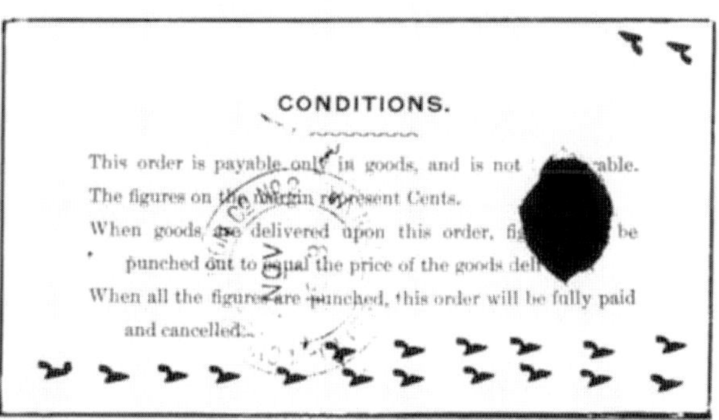

CONDITIONS.

This order is payable only in goods, and is not [illegible]able.
The figures on the margin represent Cents.
When goods are delivered upon this order, fi[illegible] be punched out to equal the price of the goods del[illegible]
When all the figures are punched, this order will be fully paid and cancelled.

Longdale Iron Company, Clifftop Mine: A drift mine on Glade Creek, nine miles north of Sewell, Clifftop opened in 1884 and operated until 1913, when Babcock Coal and Coke Company assumed ownership. Pictured above is paper scrip issued by Longdale Iron Company to Dave Pugh, dated 1898. Dave Pugh had migrated from Wales and settled in this area at that time. Mr. Pugh is the great-great-grandfather of Rebecca Reno Black of Crab Orchard, who shared the paper scrip for this publication.

SHERWOOD

SHERWOOD MINE—POWER HOUSE, TIPPLE AND DWELLING HOUSES

Price Hill Fuel Co., Fayette County

Daily capacity—1250 tons coal

Sherwood mine, powerhouse, tipple, and dwelling houses – Price Hill Fuel Company.

Sherwood – Fayette County: The New River Company was a huge enterprise requiring many highly capable people to operate it. Robert Haven Gross served the New River Company for thirty years as a top administrative leader in different capacities. Charles R. Bourland served as both president and vice president. J. A. Hunt was vice president of operations for the New River Company. Hunt also had business ventures in Lillybrook, Raleigh Smokeless, Fire Creek Smokeless Coal Company and several others. L. Ebersole Gaines was president of New River from 1939-1945. S. A. Scott came to the New River Company in 1911 as general manager and president and served until his death in 1938.

Some more recent personalities at the New River Company were W. A. "Bill" Haslam, who served New River from 1951-1970 to become president in 1960. E. V. Bowman came to New River Company in 1960 and later became president. Harry L Kirkpatrick II started working as an engineer and in 1970 became the vice president. O. T. Carr held the position of secretary-treasurer for many years.

Sherwood E. Sparks, Martin L. Garvey, Clifford Clapper, Dr. Gory Hogg, Charles E. Mahan, Thomas William Howard, Charles E. Howard, Edward J. McCool and Stanley Higgins, Jr. were all instrumental in operating the New River Company.

SKELTON

New River Company store at Skelton – Raleigh County.

Skelton mine, tipple, and powerhouse of Cranberry Fuel Company.
COURTESY ROBERT P. BODNAR

SKELTON

The superintendent's house of the Cranberry Fuel Company at Skelton. This house was dismantled and moved to the Beckley Exhibition Mine for tourist tours.

The Cranberry Fuel Company was opening two drift mines at Shelton and Sprague in June of 1905, while the railroad access was still under construction. This was a six-mile spur line to the Piney Branch of the C & O Railway. About 300 men were employed in construction and development work in opening these new mines.

Skelton was named for a large coal mining and industrial town near Samuel Dixon's birthplace in England.

SLAB FORK

View of Slab Fork from the Virginian railway bridge.
Slab Fork Coal Company.

Slab Fork – Raleigh County: Slab Fork Coal Company received its charter of incorporation in March 1907. W. Gaston Caperton moved into the wilderness in 1907 as general manager of the Slab Fork Coal Company and became president in 1928, succeeding his older brother, George Henry Caperton. The brothers worked closely together from the early 1890s. The Capertons opened the first mine in what was later to become the Winding Gulf District. At the time the only means of transportation in and out of Slab Fork was either by foot or horseback over bridal paths.

The Capertons had the distinction of loading the first railway car of coal in Raleigh County from their Winding Gulf coal fields.

In May 1907, Beaver Coal Company leased 2,671 acres to Slab Fork for coal mining purposes. Over the years, others tracts of land were leased and by 1954, Slab Fork had approximately 6,600 acres.

The annual output from this property grew from 250,000 tons in 1910 to approximately 650,000 tons in 1956.

SLAB FORK

Store and office building of the Slab Fork Coal Company.

School house.

Sectional view of miners' homes.

The Florence Theatre.

SLAB FORK

The Caperton residence at Slab Fork.

The Caperton brothers, George Henry and William Gaston, came from Fire Creek in Fayette County where they were engaged in mining. William Gaston established his residence in Slab Fork where he opened the company's first mine. George Henry moved to Charleston to devote his time principally to the sale of the company's coal. The two brothers were a perfect team and under their guidance the company grew and prospered. The coal industry remained a family tradition for four generations.

Slab Fork Coal Company received its charter of incorporation on March 9, 1907, executed its lease with the Beaver Coal Company on May 23, 1907, and shipped the first car of coal on November 30, 1907. Slab Fork mined about 62,000 tons in their first year of operation. The coal was undercut with picks, loaded by hand, hauled by mules, dumped into a bin and loaded from side chutes to railway cars.

View of Slab Fork – 1907.

View of Slab Fork Coal Company tipple.

SLAB FORK

Slab Fork – Raleigh County: When Mrs. Beulah Milam moved to Slab Fork in 1948 as a young housewife, "There were 214 houses, a post office in the company store, and a doctor's office in a separate building on the hill. In 1982 Slab Fork Coal Company closed the mines, later pulling out the equipment and other things that needed to be done to close the mines permanently. In April 1983, they closed the offices for good and Slab Fork Coal Company was no more."

"The people that wanted to remain in Slab Fork were allowed to lease their homes from Beaver Coal Company, LTD. Today (2003) there are thirty-five houses, thirteen mobile homes, one business and the post office left in the community," says Mrs. Milam.

"After the coal company moved out, the people left in Slab Fork leased the water and sewer plants and began to operate them as a community project. We continued this task until June of 2001, when our water supply ran low. On June 23, 2001, the Raleigh County Public Service District, located in Sophia, turned a valve on that started water flowing into our community, after they had spent many man hours and laid many feet of pipe to bring in water from Sophia. We are grateful to have this water and no longer have to keep up an outdated water system. The people in the community still operate the sewer system at this time," reports Mrs. Milam.

Mrs. Milam said, "Not to forget the good doctors they had over the years. Some I can't remember the first name, but we had Dr. Green, Dr. Simmons, Dr. W. F. Pomputius, Dr. Lewis Fox, Dr. Satoe, Dr. Daniel Nichols, and the wonderful company nurse of many years, Mrs. Kenneth (Wilda) Bays."

Complete set of scrip issued by the Slab Fork Coal Company.

Mine No. 2 - Slab Fork Coal Company. Tipple, railroad bridge and part of the town.

SPRAGUE

SPRAGUE MINE — TIPPLE

Cranberry Fuel Co., Raleigh County

Daily capacity—1000 tons coal

Sprague – Raleigh County: Tipple, circa 1910.

PHOTOGRAPHS OF SPRAGUE COURTESY J. W. DIXON, JR.

SPRAGUE MINE AND TOWN

Cranberry Fuel Co., Raleigh County

Daily capacity—1000 tons coal

Sprague mine and town.

As railroad construction continued in 1905, Cranberry Fuel Company was being organized to open mines at Sprague and Skelton. The Sprague mines ended operations in 1951, after having mined 11,917,528 tons of coal since 1906. The company store closed in 1954 and the coal company sold the houses. The town was named for Phineas W. Sprague, who with Samuel Dixon formed the New River Company.

STANAFORD

The Stanaford tipple of the New River Company – Raleigh County.

The New River Company purchased the mines and houses of the community of Stanaford. Mines that previously had operated under Piney Colliery Company, Elkhorn-Piney Coal Mining Company, and the Koppers Coal Company. Coal operations ceased with the closing of the mines in 1955. COURTESY ROBERT P. BODNAR

STICKNEY

Stickney – Raleigh County: Tipple, circa 1930.

The Stickney mine, six miles south of Whitesville, first operated as the Glogora Coal Company of Mordue, then as American Rollings Mills Company (AMRCO). It shipped coal to their steel plant. The last owner and operator of Stickney mine was Eagle Coal & Dock Company. COURTESY HORACE ROBINSON

The Stickney tipple pictured above was comprised of three button lines to reach the tipple from the head house. It has been said this is the longest button line in the world.

STOTESBURY

Recreational building of the E. E. White Coal Company.

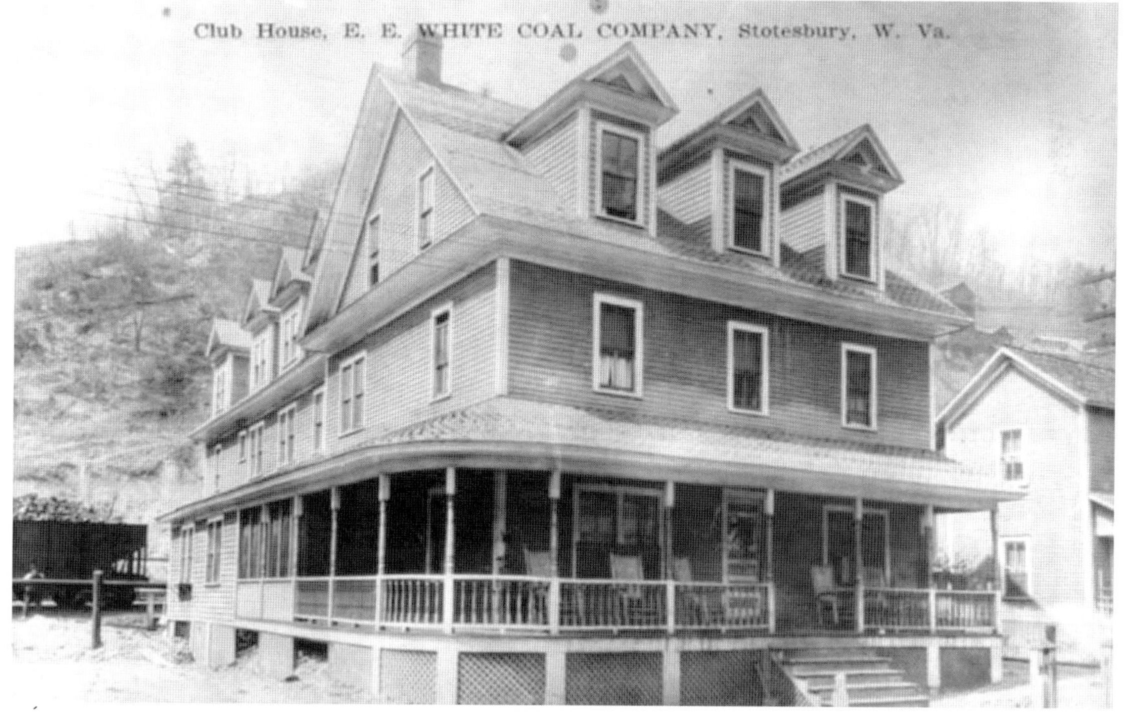

Club house.

Stotesbury – Raleigh County: Was named by the company president, E. E. White in honor of Edward T. Stotesbury, president of Beaver Coal Company. The company started with a lease of 5,000 acres and capital of $500,000.

STUART

STUART MINE—TIPPLE AND POWER HOUSE

Stuart Colliery Co., Fayette County

Daily capacity—1250 tons coal

Stuart mine, tipple, and powerhouse – Fayette County.
Stuart Colliery Company.
COURTESY J. W. DIXON, JR.

Samuel Dixon named this town Stuart because of his English background. England has a town and royal family by this name. Samuel A. Scott renamed the mine Lochgelly in 1912 because of the fear of the "gassy" mine.

In early February of 1907, a total of 89 men died in an explosion inside the mine. The dead included the bank boss, mining engineer, and fire boss. The Stuart mine was a shaft 584 feet deep and was very close to the Parral shaft mine where 23 lives were lost in an explosion in 1906.

Mr. C. C. Rose, a Beckley undertaker, worked at the scene of the explosion in Stuart, and said, "all kinds of reports have been circulated as to [sic] the cause of the explosion, but this will really never be known, as not one who was in the position to know escaped death."

Samuel Dixon named the mining town of Parral after a coal-mining town in Mexico that he visited in 1905. The new name, Summerlee, was taken from a coal town in Scotland.

See also pages on Lochgelly.

SUN

Sun – Fayette County: The Sun Coal & Coke Company started in 1893, and was one of the oldest in the New River field. Sun was first opened by John Laing (1865-1943), and his uncle James Laing (1847-1907). By 1896 they had the mines operating. Mine No. 1 was a slope with a screen tipple, while No. 2 was a shaft. Later No. 3, another slope, was opened.

Sun Coal & Coke sold out after seven years to New River Smokeless in 1900. In 1907 the mines were sold again, and renamed New River Collieries. Sun had a daily output of 1500 tons, with 160 miners employed and 130 day laborers mining the Sewell seam. Mr. W. A. Mills (father of Maude Mills Skiles) became the general mine foreman for Sun from 1912-1929.

Stonega Coal Company of Big Stone Gap, Virginia, took over a one-half interest in the Sun mines in 1923 with Chicago, Williamson & Franklin of Illinois. After a dispute between the owners, the Sun mines closed on December 1, 1931.

Sun grade school.

In front of the company store – Sun.

James Laing, born in Scotland, first settled in Pennsylvania and then moved to Quinnimont, working as a miner and later becoming manager of the mine. Mr. Laing organized other mines over the years, including Royal Coal & Coke Company near Prince, Lanark Fuel Company, and the Sun Coal Company. These companies were consolidated in 1906 with the New River Smokeless Coal Company and the Red Ash Company.

The boarding house at Sun in the early 1900s. The men pictured left to right are (last names only) Gores, Ferguson, Chase, Brand, Stoneburner, Hogg, Casey, Jones, Jones, and Craig. PHOTOGRAPHS OF SUN COURTESY FRED RISK

TAMROY

Tamroy tipple – McKell Coal & Coke Company.
COURTESY E. R. REYNOLDS COLLECTION

Tamroy – Fayette County: Was named by William McKell for a town in his native home in Scotland. The mines were located on Dun Loup Creek. Mr. McKell built his own railroad, Kanawha, Glen Jean & Eastern Railroad, to serve his mines of Tamroy and Oswald. The New River Company purchased all McKell properties in 1940.

TAMROY

A commemorative envelope from the Tamroy-Oswald reunion in 1989. Jeannie Kincaid's son, Larry Ray Kincaid, did the artwork.

The family home of Mr. Ralph Wells located on South Sand Branch, near Prosperity. Pictured are his parents, Creed and Agnes Wells, his wife Wandalene, and daughter Jean.

TAMS

Tams tipple, powerhouse, and railway station – Raleigh County.
Gulf Smokeless Coal Company.
E. R. REYNOLDS COLLECTION

The Tams mine, under the direct management of Mr. William T. Tams of Scranton, Pennsylvania, was the largest single producer in the state in 1911. Loading and shipping was as high as 2,000 tons per working day.

Major W. R. Tams left the New River Company in 1908 to form the Gulf Smokeless Coal Company at Tams in Raleigh County. He sold his holding to Winding Gulf Coal in 1955, after building a record that was unmatched in the coal industry.

Construction of the town of Tams got under way in May 1909 and within a year 125 houses had been completed. The store and office building were completed in 1910.

Tams had a theatre, rows of neat houses, which were freshly painted each year, sidewalks with picket fences, a schoolhouse, churches, and a huge company store. Major Tams designed a brick bathhouse for the use of the miners.

The theatre building featured a room for dancing, a gymnasium with a basketball court, bowling alley, pool tables, and a small café.

Carload of Beckley coal at Tams.

Train station and tipple – Gulf Smokeless Coal Company.

THURMOND

Thurmond – Fayette County: After the Civil War, Mr. Morris Harvey acquired land in Fayette County. He hired Capt. William D. Thurmond, a self-taught surveyor, to make some surveys of his extensive holdings. Thurmond had to take his pay in land of his selection. Deciding that the land on which Thurmond would be built would become the great port of commercial entry into the Fayette County coal field, Captain Thurmond asked for that location. When the C & O Railway built its line there, the place took on great importance as a depot, post office, and business site. Thurmond is situated at the mouth of Dun Loup Creek, along the banks of the New River. The business section lies crowded in the narrow confines between the river's bank and the mountains.

The Dog Wagon at Thurmond Depot – 1916.
Man at left is Charley Ash, other man unknown.
PHOTORAPHS OF THURMOND COURTESY FRED FRISK

The Mankin Building was constructed in 1904 and was the home of the New River Banking & Trust Company and of Mankin Drug. A dentist's office was located on the third floor. The New River Bank opened its doors August 11, 1904. The bank moved to Oak Hill in 1935.

THURMOND

Thurmond depot.

The nearest thing Thurmond had to a turnpike or road was the main line of the C & O, which ran through town. In its peak period there were 14 passenger trains that rolled daily through Thurmond. It was one of the busiest places in the New River Gorge area. Within easy traveling distance, there were 67 coal mines, large and small. There also were 22 train crews and 12 yard crews working and living in Thurmond.

Dun Glen Hotel, built by William G. McKell in 1901. Destroyed by fire July 1930.

Businessmen coming into Thurmond, early 1900.

TOMMY CREEK

Tipple – Tommy Creek Coal Company – Raleigh County.

Company store and office building.

TRACOAL

Tracoal – Raleigh County: Company store and office – Trace Fork Coal Company.

General view of Tracoal.

TRALEE

Tralee – Wyoming County: Store and office - Barkers Creek Coal Company.

Around 1910 Mr. J. C. Sullivan opened a mine on the hill above the public road south of Mullens, and called it Mead Poca. Mr. Sullivan was also one of the owners and managers of Harty Coal Company, which went into operations on a lease acquired in 1915. Sullivan named the post office in honor of his mother.

Tipple and Incline,
BARKERS CREEK COAL CO.,
Tralee, W. Va.

Tipple and incline, Barkers Creek.

Company store – Meade Pocahontas Coal Company.

Amusement hall - Harty Coal Company.

J. C. Sullivan interests.

General office building.

President's residence.

VANWOOD

Vanwood – Raleigh County: Miners' homes – Wood-Sullivan Coal Company.

WHIPPLE

View of Whipple – Fayette County.

Whipple Colliery Company.

WHIPPLE

Whipple mine was named in honor of William Whipple of Portsmouth, NH, and is located near the corporate limits of Oak Hill. Mr. Whipple acquired the property, located the shaft, and developed the Whipple mine. Whipple was one of a group of six producers of the famous White Oak Smokeless coal.

Whipple is located on White Oak creek and began production in 1900. It closed in 1957.

Whipple mine – powerhouse and tipple – White Oak Fuel Company.
COURTESY J. W. DIXON, JR.

Tipple at Whitby – Raleigh County.
Spencer Fork Coal Company.

Miners' homes- Spencer Fork Coal Company.

Residences of general manager and superintendent.

WICKHAM

Wickham – Raleigh County: In the early days this community was known as Bickel and the "Beckley Slope." Later it was named for coal operator Thomas Wickham who had purchased the land from W. R. Graham, of Tazewell, VA.

Beckley Coal & Coke Company of Wickham was organized in 1903 under the administration of E. E. White and Tobias Bickel. It was a drift mine located on the Piney Branch of the C & O Railway. The Sewell seam was mined and a double entry system was used.

In 1906 Samuel Dixon of the New River Company bought both the Beckley Coal & Coke and Mabscott Coal & Coke operations. At the time of this purchase, Beckley Coal & Coke Company's daily output of coal was 1,250 tons.

MINERS' HOUSES — TIPPLE AND GENERAL STORE IN THE BACKGROUND
Beckley Coal and Coke Company, Raleigh County

Miners' houses, tipple, and general store in the background.
Beckley Coal & Coke Company.
COURTESY J. W. DIXON, JR.

View of Wickham.

Tipple and mines – Beckley Coal & Coke Company.

THE NEW RIVER CO. 25M-6-14-(14) 33684.

No. 73 SEP 26 1914, 191....

J. B. Kirk

IN ACCOUNT WITH Beckley Coal & Coke Company

Pay Roll Clerk Stamp Name of Company in Space Above.

For Month of SEP 15 1914 191....

CR.		DR.	
By 1354 Cwt. @	27 08	To Store	8 00
By ...... Cwt. @		To Overdrawn Last Month	
By ...... Cwt. @		To Cash	
By ...... Hours @		To Rent	
By ...... Hours @		To Electric Lights	
By 6/9 Hours @	1 50	To Coal	
By ...... Yards @		To Hauling	
By ...... Yards @		To Smithing	25
By 8 Yards @ 50	4 00	To Insurance	75
By ...... Board or Labor		To Workmen's Comp. Fund	05
By ...... Cash		To Doctor	50
Insurance	30 00	To Mine Checks	
		To Board or Labor	
Total	62 58	Total	9 05
To Balance Due Company		By Balance Due Workman	53 53

E. & O. E. Liable for any Subsequent Indebtedness.
Retain This Statement.

The New River Company employee statement.

WICKHAM

Mr. Howard Shrewsbury at the Wickham Mine.

From the 1919 Report of WV Department of Mines:

Wages: Run of mine, $1.08 per car
Pick miners paid per ton of 2240
Selling price of coal, $2.79
Machine runners paid $5.40 per day
Machine helpers paid $4.45 per day
Number of men killed, 9
Number of mine openings, 22
Acres of coal worked-out, 138
Number of men injured non-fatally, 25
Number of horses and mules used at the mines, 8
Number of fans, 22

No. 34

Mabscott, W. Va., 9-13 1919

ROAD TAX, TOWN OF MABSCOTT 1918

Received of J. B. Kirk

One Dollars

$ 1.00

A L Kelly

Sergeant

Form Prescribed by Department of State Tax Commissioner for Inspection and Supervision of Public Offices, etc.

No. 1121 Dec. 3 1914

Received of Kirk J. B.

The Taxes assessed for the year 1914 on the following described property:

Town of Mab (Town Dist.) Raleigh County, W. Va.

Tax Levied on Each $100 Valuation	
State Levy	10c
County Levy	20c
District Teachers Fund	25c
District Building Fund	12½c
Total	67½

Estate Held	Acres	DESCRIPTION	VALUATION	TOTAL TAX
D.		East Mickham	300	2 18
			75	50
				2 57
				2 57

G. W. Thompson

Sheriff Raleigh County.

M. D. Pugh

To R. H. DICKINSON, Sheriff of Fayette County, W. Va., Dr.

FAYETTEVILLE DISTRICT

1910. Tax Levied on each $100.

State Tax	4½c
State Road Tax	1c
County Tax	20c
Road Fund	10c
Teachers' Fund	25c
Building Fund	12½c
Special Levy	10c
Total	

Lot #6 Acres

LOCATION Scarbro

Valued at $ 50.00

Total Tax 4.65

10 per ct. int. from Jan. 1

2 1-2 per ct. discount

S. F. C.

Received Payment W. A. Edler

Fayetteville District.

1910 Real Estate.

NAME

Amount of Tax

Interest

Discount

Amount Collected

Beckley, W. Va. 6/30 1913

Mr. J. B. Kirk

To BECKLEY ELECTRIC LIGHT & POWER CO., Dr.

To Balance Lights

" " Water

" " Wiring, Lamps, Etc.

" Water 191

" Lights 191

" Steam 191

" Meter Reading Lights 137

" Meter Reading Lights 131

6 Kw. Hrs. @ 10c 60

" Meter Reading Water

" Meter Reading Water

Gal. @

To

Received Payment June 30 1913 $ 60

A. E. Harvey Collector.

Receipts from miner, J. B. Kirk.

The powerhouse of the Beckley Coal & Coke Company.

WINDING GULF

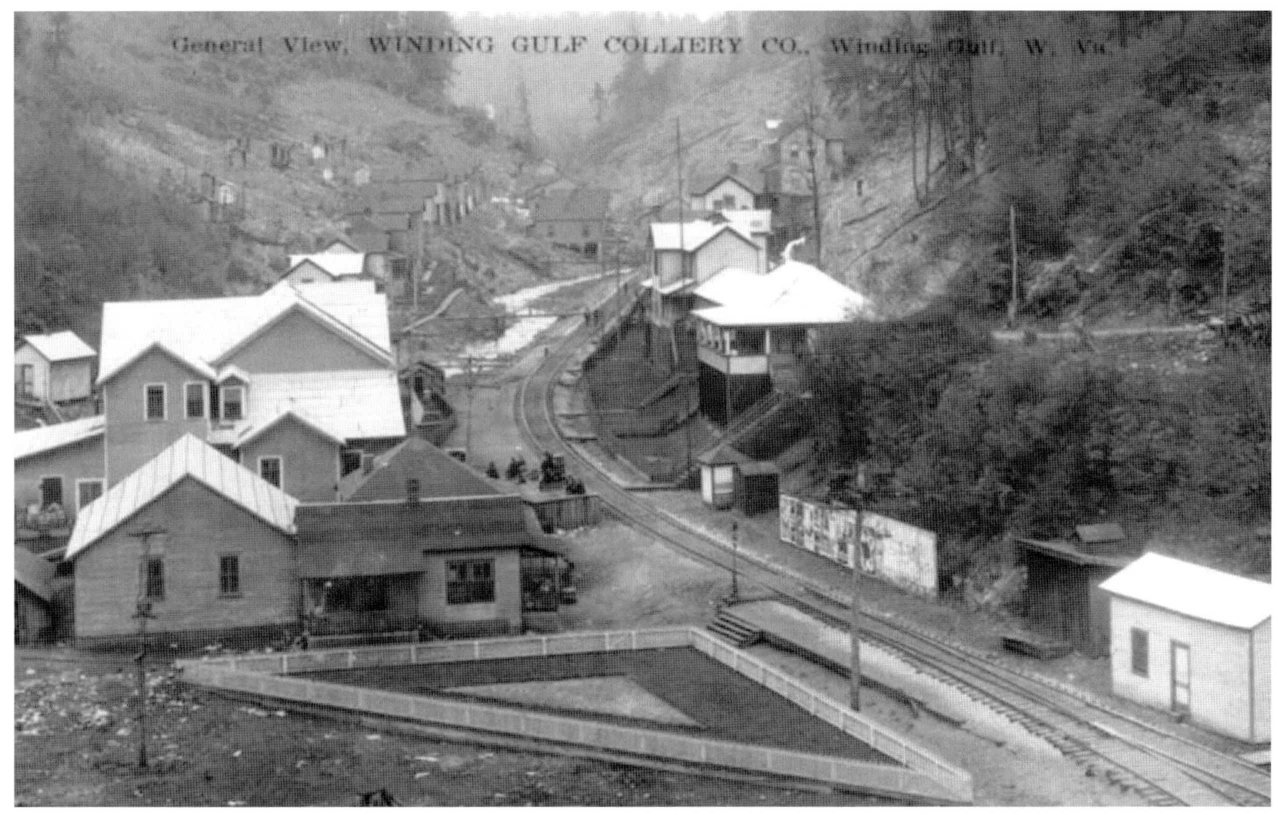

Winding Gulf – Raleigh County: General view of Winding Gulf Colliery Company.

Miners' homes, Winding Gulf.

WINDING GULF

Winding Gulf Colliery Company - tipple, store, and powerhouse.

Justus Collins, owner and founder of Winding Gulf Colliery Company, had extensive holdings in Fayette County before he started to develop the area of Winding Gulf. He donated land for the junior high school in Oak Hill that bears his name.

Tipple and powerhouse – Winding Gulf Colliery Company.

Marbles for the young and old.

High school children – Winding Gulf Colliery Company – Winding Gulf.

WINDING GULF

Church and school – Winding Gulf.

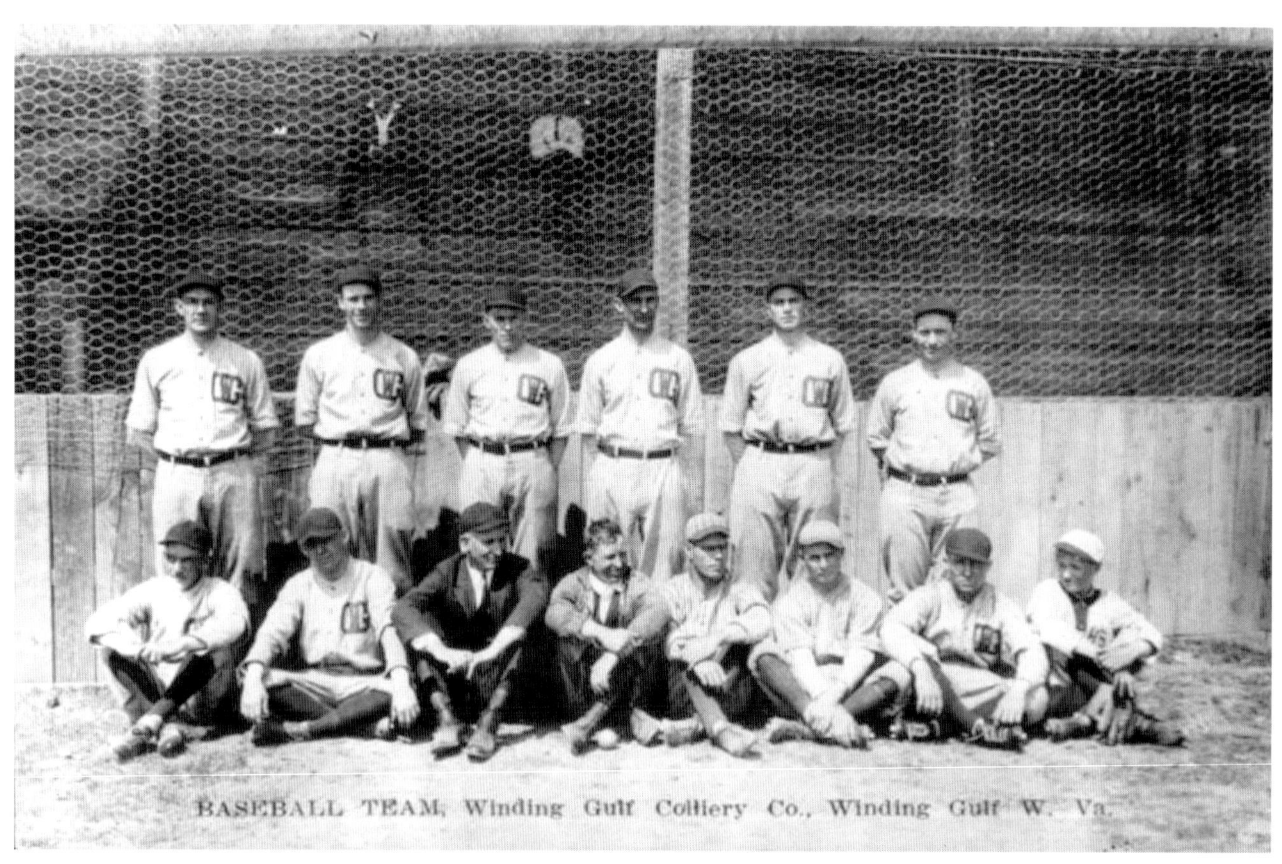

Winding Gulf Colliery Company's baseball team.

WINGROVE

During construction of Wingrove.

Wingrove – Fayette County: A pioneer settler, John Wingrove built a log cabin and set about making a home for his family. The soil was fertile and game was plentiful. Adam Blake became a neighbor to Wingrove and he built a log house for his family. A little hamlet near the Wingrove farm took its name from the valley's pioneer family.

White Oak Fuel Company operated Wingrove and according to the *1908 Department of Mines Report*, Wingrove employed a total of 132 workers. The average number of working days that year was 234, with 28 horses or mules. White Oak Fuel Company used Scarbro as its address. This 1912 report showed White Oak operating Wingrove, Scarbro, Whipple, Carlisle and Oakwood.

INDEX

BIBLIOGRAPHY

Amick, Dorothy D., *The Methodist Temple of Beckley, WV.* Jarrett Printing Col, Charleston, WV, 1966.

Dillon, Lacy A., *They Died for King Coal.* Apollo Books, Inc., Winona, MN, 1985.

Feller, John W. "Jack". *Mullens, West Virginia.* Harless Printing Company, Inc., St. Albans, WV, 1993.

Morgan, John C., *West Virginia Governors.* Charleston Newspapers, 1980.

Peters, J.T. & Carden, H.B. *History of Fayette County, West Virginia.* Charleston, WV, 1926.

Southern Bankshares, *1989 Annual Report of Beckley National Bank.*

Tams, W.P. Jr., *The Smokeless Coal Fields of WV: A Brief History.* West Virginia University Press, Morgantown, WV. 1963.

U.S. Dept. of Labor, *Historical Summary of Mine Disasters in the U.S. Volume I,* Coal Mines 1810-1958. 1998.

Various articles from the *Beckley Post Herald* and *Raleigh Register.*

West Virginia, *Annual Report of the Department of Mines,* 1895-1933.

Warren, Harlow, *Beckley USA, Volumes I, II & III,* 1955, 1963, 1968.

Wood, Jim, *Raleigh County, West Virginia.* BJW Printers, Beckley, WV, 1994.

Daniel and Mary Stevenson with Governor Cecil H. Underwood. The occasion was a tea in the Governor's Mansion to honor the West Virginia History Heroes for 2000. Mary was a recipient that year. *Courtesy John W. "Jack" Feller*